THE BOOK OF NUMBERS

THE SECRET OF NUMBERS AND HOW THEY CHANGED THE WORLD

PETER J. BENTLEY

THE BOOK OF NUMBERS

THE SECRET OF NUMBERS AND HOW THEY CHANGED THE WORLD

FIREFLY BOOKS

A FIREFLY BOOK

Published by Firefly Books Ltd. 2008

Second printing, 2008

Publisher Cataloging-in-Publication Data (U.S.)

Bentley, Peter J.
   The book of numbers / Peter J. Bentley. [272] p. : col. photos. ; cm.
Includes index.
Summary: Explores the fascination with numbers and mathematics in every aspect of life from The DaVinci Code to the Enigma Code, as well as in the digital world.
ISBN-13: 978-1-55407-361-0
ISBN-10: 1-55407-361-8
   1. Mathematics — Popular works. 2. Number theory — Popular works. I. Title.
510 dc22   QA93.B468 2008

Library and Archives Canada Cataloguing in Publication

Bentley, Peter J., 1972–
   The book of numbers / Peter Bentley.
Includes index.
ISBN-13: 978-1-55407-361-0
ISBN-10: 1-55407-361-8
   1. Mathematics — Popular works. 2. Number theory — Popular works. I. Title.
QA93.B44 2008   510   C2007-904540-5

Published in the United States by
Firefly Books (U.S.) Inc.
P.O. Box 1338, Ellicott Station
Buffalo, New York 14205

Published in Canada by
Firefly Books Ltd.
66 Leek Crescent
Richmond Hill, Ontario L4B 1H1

For Cassell Illustrated:
Publishing Director: Iain MacGregor
Commissioning Editor: Laura Price
Editor: Jenny Doubt
Creative Director: Geoff Fennell
Layout: Keith Williams
Production: Caroline Alberti

Cover photography: Corbis/Matthias Kulka/zefa back top left; /Joseph Sohm, Visions of America back bottom left; /Gustavo Tomsich back centre. Getty/Jeff Foott back top right. Science Photo Library back bottom left.

Printed in China

CONTENTS

Numbers flow past us like a blizzard, wherever we are on the planet. We drive in rivers of numbers. We listen to numbers on headphones. We wear changing numbers on our wrists.

BEFORE THE

CHAPTER -1

We live in numbers, talk in numbers and watch numbers for entertainment. Numbers rule our lives, they wake us up, tell us where to go, how to get there and when to leave. Numbers are judges of all, they assess and compare with complete authority and impartiality. But numbers also lie; they may mean anything except the truth. Numbers can save our lives, and a love of the wrong kind of numbers can ruin us. Numbers can be our friends, our lifelines and our lucky charms. Numbers can also kill us. You are made of numbers. So am I.

Thousands of years ago, when there was no difference between science and religion, numbers seemed to hold the key to understanding the universe. They might not have trickled down in front of our eyes like a scene from *The Matrix*, but lurking inside many forms were important numbers that seemed too common to be coincidental. The same ratios kept reappearing in nature, perhaps between the diameter of a circle and its circumference, or in the curvature of seashells. The same geometric shapes and the numbers embedded within them kept being discovered in unlikely places, such as the spacing of the planets of our solar system. Even something as unlikely as a speed – for example the speed of light — seemed to be at the heart of the construction of our universe. In those days it was widely believed that such numbers pointed to the mysterious underlying design of God. An understanding of those numbers would be like reading divine messages written into the fabric of existence. Those pioneers and adventurers who dared to explore the uncharted territories of numbers were exploring the very substance of their world. They were unpicking the details of life, the universe and everything. Their solutions were not a single number, but a whole collection of important numbers, as well as tools to manipulate those numbers.

Today, science has taken over from religion. We still believe that our universe has hugely important numbers associated with it. We now know that they are the patterns visible in the woven tapestry from which everything is made. Some patterns in the tapestry are made from such thick threads that they catch the eye immediately: numbers such as π, e and θ. Some form the bulk of the material: numbers like 0, 1, 2, 3 and $\sqrt{2}$. Some stand out like accidental spills on the fabric, such as 10 and 13. Other numbers and concepts, like c and ∞, point to the size and shape of the tapestry. Some, such as i, are only visible as tenuous ripples of complexity that flow through the cloth.

BEGINNING

Those who explore the fundamental truths of nature are now known by names such as mathematicians, astronomers and physicists. But however we label them, these people were (and still are) explorers. They did not weave the tapestry they examined. They did not invent their numbers or mathematical ideas as a novelist might invent a story. They searched for the truth, and tried to explain it, inventing new languages of numbers simply to be able to write down their discoveries. Some pursued this goal for science, some for religion and some for fame.

The explorers we will follow in this book were very clever and many have been called geniuses — but they were people too. They had complicated lives, had arguments, failings and successes. Galileo was a medical school drop-out, Newton threatened to burn down his parents' house, Bernoulli stole his son's work, Pascal was a bully and Einstein had a child out of wedlock. Some were murdered because of numbers. Others lost their sanity. Put them all in a room together and you'd probably be deafened by the shouting. But they all had an appreciation of numbers that made them exceptional. They came from all over the world, yet their language was universal. As their explanations improved, so did the language of numbers.

Through our quirky pioneers we learned how numbers make shapes, angles and connections, enabling us to measure our land, to design and build complicated machines. We discovered the numbers of interacting waves, allowing us to understand music, the swing of pendulums and the bizarre properties of light. We learned how numbers describe position, speed and accelerations, enabling us to understand the motion of the planets and comprehend the planet we live on. We learned how numbers define time, space and the different sizes of infinity, enabling us to understand how the flow of time changes and how our universe began. Today we continue to learn the numbers that affect subatomic particles, and those behind complex systems such as economies, societies and consciousness. These remarkable achievements have created our modern world of telephones, cars, music, computers and airplanes. They have enabled almost every modern device you use, the food you eat and the job you do. Your entire lifestyle is shaped by our understanding of numbers.

This book is about the explorers of numbers and the inventors of mathematics. The motivations and beliefs of these eccentric characters are often surprising. However, more surprising than their discoverers, are the numbers themselves.

Albert Einstein once said, "There are two ways to live your life. One is as though nothing is a miracle. The other is as though everything is a miracle." Numbers don't remove your ability to marvel at the world, they increase it.

Numbers are miraculous, as you're about to find out.

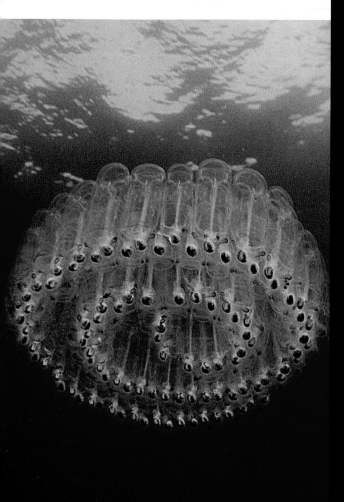

As far as we know, humans are the only creatures on the planet to recognize and manipulate numbers. We can train a parrot to count or a dog to do simple arithmetic, but they do not naturally have these abilities. But does that mean that numbers need us to exist? That without us there would be no numbers, no counting, nothing? What are numbers, anyway?

MUCH ADO ABOUT

CHAPTER 0

Numbers are words (and symbols) that we use to describe patterns. It is essential for all creatures on our planet to be able to perceive patterns. Even the simplest organism needs to be able to distinguish between things that may kill it and things that it needs to eat. More complex organisms need to be able to distinguish between *less* and *more* food. Animals that rear young need to have a very good intuition about whether all their young are present or not. Other animals need to know the difference between two spots of light that might be the eyes of a predator and several spots of light that might be camouflage or random reflections. So many creatures, including humans, have evolved brains that are naturally very good at spotting and distinguishing between patterns of things.

The numbers we write down and speak are words in a language, and that language is called *mathematics*. We are the only creatures on Earth to make use of language, so it is not terribly surprising that we are the only ones to "speak numbers." But patterns will always exist whether we name them or not. We happen to call one pattern of similar objects "three," and another "four." But naming them, or even counting them, does not change the number of them. Some people ask whether a handclap in a forest that is heard by nobody actually makes a sound. Of course it does – sound does not require ears, it is a vibration of molecules. Similarly, a number (or a pattern) that is seen by no one is still a number, whether we are there or not.

Writing Numbers

Numbers have been our friends for thousands of years. Although we may have discovered them at the same time as we invented the flint hand-axe, numbers have taken a long time to grow into the forms we are familiar with today. Numbers weren't created overnight. No genius caveman (or cavewoman) woke up one morning, grabbed a stalagmite and started scratching "1, 2, 3" in the dust. No, numbers started life as unrecognized, unspoken, unnamed ghosts. As we've changed and developed, so they've grown into the solid forms that now rule the world.

NOTHING

Right: The number of spots on a ladybug denotes its species and may work as a defensive pattern against predators.

Many thousands of years ago, when people didn't speak many words, before writing was invented, before there was money, before there were even words to describe numbers, people knew numbers. Although we had no names for them, we used them. We couldn't think about them or draw them. We could only tell the difference between one, two, three and many things, rather like being colorblind to quantities. If you lived at that time – no matter how intelligent you were – you would have struggled to tell the difference between six apples and seven, just by looking. Even though you had the same eyes and the same brain, you would have found it hard to tell the difference. Why? Because counting hadn't been invented.

Counting is a really clever trick to know – especially if you have no idea what numbers are. The first people to count may have seemed like magicians or shamen. Their magical counting abilities were perhaps first needed when tribes

Below: Mexican Mayan glyphs relating the transfer of power from Pacal I. Numbers are represented by human bodies. This comes from a relief palace panel dating from 702 A.D.

of people first began fighting each other (a long, long time ago). If you're the chief of the tribe and you send a large number of your warriors out to defend (or attack) another tribe, it's pretty useful to know whether they've all come back or not. And some tribes had a tradition of demanding reparations in proportion to the losses (I lost 15 men, so I want 15 buffalo from you). But if you have no word for 15, and no method of counting, how can you ever achieve a fair compensation?

The trick they used was very simple: as warriors left to go to battle, they each put a rock in a pile. When they got home again, they each removed one. The number of rocks left equaled the number of men lost. The chief would then remove each rock in turn and pick up a stick for each rock he removed (sticks are easier to carry). He then would walk to the other tribe with his sticks and demand a buffalo in exchange for each one. So without actually counting or even understanding the notion of numbers, it was possible for very precise trades and transactions to take place.

The trouble with using rocks and sticks is of course that they take up space and may be lost. A bucket full of rocks or a sack full of sticks is one way of "writing down" a number, but it's not the most efficient method. (In fact the storage in special pots of clay pellets shaped into different tokens was used 6,000 years ago in Elam, in present-day Iran near the Persian Gulf.)

But we've been writing numbers in a more efficient way for 30,000 years. The clues are visible in animal bones that have been found marked with intricate notches. In prehistoric

Above: A system of counting on fingers, from folio 1V of De Numeris, *a ninth-century manuscript of* Codex Alcobacense *by the German theologian and writer, Raban Maur (780–856).*

times, people would use their flint axes and cut grooves, so they could keep records of numbers. One notch a day measures the passage of time and allows you to predict lunar cycles or seasons with great accuracy. One notch per animal allowed the first shepherds to know if all the flock were still there at the end of the day. One notch per kill allowed the tribe's best hunters to prove their bravery and skill. Interestingly, the notches were often made in groups of five. This was done for two reasons — first because we happen to have five fingers and we've been using fingers to count with for as long as we could move them independently. But the notches were grouped into fives for another reason — as we've already seen, the human

Above: Another image of a
finger-counting system from the
ninth-century manuscript of
Codex Alcobacense by Raban
Maur (780–856).

The very same method was used by the Romans, many tens of thousands of years later. It's no coincidence that Roman numerals were written down as I, II, III, IV, V, and so on — their origins were in the grooves cut into bones or wood, tens of thousands of years before. The Romans used a "V" for the number 5 for the same reason that prehistoric man grouped his notches in fives — it's much easier to understand V at a glance compared to IIIII. Clues of the origins of the Roman numerals are also visible in their language. In Latin, *rationem putare* means "to count." The word *ratio* meant (as it does today) "a relationship between things" and the word *putare* meant "to cut, or prune a tree." So when the Romans talked about counting, they spoke words that effectively meant, "use your eyes to observe relationships between things and make cuts in wood."

brain is not very good at distinguishing between two sets of many objects at a glance. So telling the difference between four, five and six notches in a row is very hard to do — and nearly impossible if you can't count. By grouping notches into five (easily done with five fingers on a hand), it suddenly becomes straightforward to see the number written down.

Despite their ancient origins, Roman numerals are certainly not forgotten. In almost every book you read today, you'll find the first few pages numbered in the same way a Roman would have counted, 2,000 years ago. Many buildings have the date they were built written in Roman numerals inscribed into a cornerstone, for example 2007 would be written as MMVII.

The ancient idea of using notches to count with is also not forgotten. Tally charts are still used today by those needing to count rapidly. Vertical strokes are made in groups of five, the fifth crossing the first four in order to make an easily readable group. Astonishingly, despite our computerized world, we still use a method of counting that a prehistoric caveman would recognize.

*Right: A Zuñi tribesman
in ceremonial dress.*

Speaking Numbers

As the symbols that we used to write down numbers with slowly developed, so did the noises we uttered when we saw them. It's possible that in the days when people lived in caves, their word for five was as simple as, "uh, uh, uh, uh, uh." But clearly this is not the cleverest way to speak a number — especially if it's a big number and the listener can't count. The solution was clear: we needed to make a different noise for each number. In many tribes around the world, where people counted mainly by using their fingers (and sometimes all kinds of other body parts), the words they used for the numbers they wanted to express were mainly to do with fingers and hands. See the Native American Zuñi vocabulary (right).

Over time, humans became more numerous; we formed villages and towns and began trading with each other, and so the need for numbers grew. It became necessary to have shorter, easier words for numbers that could be said a little quicker than, "all the fingers and one more raised of jugs of milk in exchange for all the fingers raised and two brought together and raised with the others of eggs." Around 4,000 years ago, several tribes developed shorter spoken words for the numbers. They never knew just how popular those words would become. Amazingly, the words that these farmers and hunters created form the basis for the words used throughout Europe today, including in Punjabi, Hindi, ancient Persian, Afghan, Lycian, Greek, Latin, German, Armenian, Italian, Spanish, Portuguese, French, Romanian, Sardinian, Dalmatian, Welsh, Cornish, Erse, Manx, Scots Gaelic, Dutch, Friesian, Anglo-Saxon and English.

Zuñi pronunciation

1	töpinte	"taken to begin"
2	kwilli	"raised with the previous"
3	kha'i	"the finger that divide equally"
4	awite	"all fingers raised bar one'
5	öpte	"the scored one"
6	topalïk'ye	"another added to what is counted already"
7	kwillik'ya	"two brought together and raised with the others"
8	khailïk'ya	"three brought together and raised with the others"
9	tenalïk'ya	"all bar one raised with the others"
10	ästem'thila	"all the fingers"
11	ästem'thila topayä'thl'tona	"all the fingers and one more raised"

Left: A 2,000-year-old series of petroglyphs representing the Fremont, Anasazi, Navajo and Anglo cultures.

Below, opposite: A relief sculpture showing the Imperial Roman Market Timetable, which uses Roman numerals.

These people are now known as the Indo-Europeans. They helped create one of the largest family of languages that we know of today — stemmed from a single original people several thousand years ago. Although we don't know exactly where these people lived, by studying the commonalities in languages, we can get a pretty good idea of how they used to say their numbers. The box shows our best guess. If you know a few languages you should be able to see how several thousand years, a thousand accents and a hundred languages have changed the words from these original "parent" versions.

Although some of the words may look a little unfamiliar, you may be surprised at how many are recognizable as numbers — especially when you remember how much other words have changed in Europe's various languages. Despite wars among European countries, the similar words we share for our numbers prove that our roots were in the same place. Our numbers bind us all together in a common history.

Parent versions

1 *oi-no, *oi-ko, *oi-wo

2 *dwō, *dwu, *dwoi

3 *tri, *treyes, *tisores

4 *kwetwores, *kwetesres, *kwetwor

5 *pénkwe, *kwenkwe

6 *seks, *sweks

7 *septm

8 *októ, *oktu

9 *néwn

10 *dékm

The Invention of Nothing

Although we could now write numbers down and say them concisely, we had not thought of everything. Or to be more precise, we had not thought of nothing. Zero had not been invented — and that was a big problem.

To see why, just imagine trying to do some basic subtraction using Roman numerals. Try:

$$\overline{LXXXIV} - DCCLIII = \overline{LXXVIIIICCLI}$$

Even when you know what the letters mean, this is not an easy calculation to make. In Roman numerals:

I = 1   V = 5   X = 10   L = 50

C = 100   D = 500   M = 1,000

And a line over a number means "multiply by 1,000." When writing numbers in Roman numerals, the symbols are normally written in order of size, with largest valued symbols to the left of smaller valued ones. If you write a smaller valued symbol left of a larger valued one, it means "subtract the value." So VI is 6, but IV is 4.

But the same calculation written in Arabic numerals seems so much easier:

$$\begin{array}{r} 80,004 \\ -\ 753 \\ \hline = 79,251 \end{array}$$

The reason this calculation seems simpler is because we use the position of numerals to give extra meaning. The number furthest to the right always means "a value less than 10." A number to the left of that always means "a number of tens less than 100." A number to the left of that always means "a number of hundreds less than 1,000," and so on. So we understand 753 to mean three and five tens and seven hundreds — or seven hundred and fifty three.

To the Romans this would be bizarre. For them, the numeral C always meant 100. It didn't change its value depending on where it was. Likewise, the numeral L always meant 50. Roman numerals did not use position in the same way that we do — which is why we can line up our numbers and do arithmetic easily, but the Romans needed to rely on abacuses.

Positional systems of numbers are great. They are so much easier to use and understand. But they do have a drawback. How do you write 10? If we no longer have one symbol with this value (e.g., X), then we have to use position to tell us the magnitude of the number. But if there are no values less than 10 to put to the right of the symbol for one, what do we put there? Remarkably, it took thousands of years before anyone worked out that we needed a new number. We needed zero.

Nothing was invented around 1,800 years ago in India. It was a very important nothing. Although the Babylonians, Greeks, Mayans and the Chinese had previously realized the need for a special symbol to be used as a placeholder to ensure the other numbers were kept in the right positions, only in India did they realize that zero meant more than that. They realized that zero is actually a number.

Perhaps the first written investigation of zero was made in 628 A.D. by a 30-year-old Indian mathematician named Brahmagupta. He was a well-respected man who was to become head of the astronomical observatory at Ujjain. Brahmagupta wrote a book entitled *Brahmasphutasiddhanta* ("The opening of the universe"), which explained the movement of the planets and how their precise paths could be calculated. By this time, zero was understood to be necessary as a positional

Above: The astronomical observatory at Ujjain in Madhya Pradesh, India, once headed by the mathematician Brahmagupta.

placeholder, to make sure the other numbers lined up properly. But Brahmagupta went a bit further. In his book, for the first time in known history, he actually defined what zero was. He said: "Zero is the result of subtracting a number from itself."

This may seem completely obvious, but 1,400 years ago these ideas were not widely understood — and not understood at all outside India. To most people, if you had, say three eggs, and you took away three eggs, then there was no result — for there was nothing left. Brahmagupta actually named that nothing. He called it "zero" and claimed that zero is actually a number. To prove it, he wrote down a series of mathematical rules to show what you do with a zero:

When zero is added to a number or subtracted from a number, the number remains unchanged; and a number multiplied by zero becomes zero.

Today a child is taught this at school at an early age, but in 628 A.D. it took the genius of a mathematician to think in this unique way. Brahmagupta also realized that zero would affect positive and negative numbers, which were usually referred to as fortunes and debts. If we pretend our debt is, say, -7 and our fortune is, +7, we can see exactly what he was saying.

In other words, poor Brahmagupta couldn't really figure out what to do with division and zeros. He didn't know what something divided by zero meant, nor was he sure what zero divided by something meant. And he thought zero divided by zero equaled zero — which was plain wrong. (If you don't believe me, try putting the three problems above into your electronic calculator and see what it says.)

But he wasn't being stupid. Other eminent mathematicians continued to follow his rules for centuries. Nobody spotted that there was a serious problem until just a few hundred years ago. It took a while, but eventually it was realized that zero doesn't always obey all the rules that other numbers obey. Take the first problem: $7 \div 0$ for instance. What could that possibly be? Well, $7 \div 2$ is 3.5. In other words, half of 7 is 3.5. $7 \div 1$ is 7 for there are seven ones in seven. $7 \div 0.5$ is 14 for there are two halves in every one and seven ones in seven.

Brahmagupta's view of zero

"A debt minus zero is a debt."	$-7 - 0 = -7$
"A fortune minus zero is a fortune."	$7 - 0 = 7$
"Zero minus zero is a zero."	$0 - 0 = 0$
"A debt subtracted from zero is a fortune."	$0 - -7 = 7$
"A fortune subtracted from zero is a debt."	$0 - 7 = -7$
"The product of zero multiplied by a debt or fortune is zero."	$0 \times -7 = 0$ and $0 \times 7 = 0$
"The product of zero multiplied by zero is zero."	$0 \times 0 = 0$

But, as clever as he was, he got a bit confused by division and zero. All he could think of was:

"Positive or negative numbers when divided by zero result in a fraction with the zero as denominator."

$$7 \div 0 = \frac{7}{0} \text{ and } -7 \div 0 = -\frac{7}{0}$$

"Zero divided by negative or positive numbers is either zero or is expressed as a fraction with zero as numerator and the finite quantity as denominator."

$$0 \div 7 = 0 \text{ or } \frac{0}{7}$$

| "Zero divided by zero is zero." | $0 \div 0 = 0$ |

So the smaller the number you divide by, the larger the result is. By that argument, 7 ÷ 0 should be infinity — you can divide 7 into an infinite number of pieces, each zero in size. This was the argument of another Indian mathematician called Bhaskara, who was born 486 years after Brahmagupta. Bhaskara (or Bhaskaracharya — "Bhaskara the Teacher" — as he is known in India) followed in the footsteps of Brahmagupta, and was also head of the astronomical observatory at Ujjain, which by then had become the most important center of mathematics in India. He wrote many books on mathematics and was the first to study and solve new types of equations. He also wrote what was apparently considered to be poetry, for example:

O girl! Out of a group of swans,
⅞ times the square root of the number are playing on the shore of a tank.
The two remaining ones are playing with amorous fight, in the water.
What is the total number of swans?

But despite his brilliance at mathematics and his dubious poetry, his solution to dividing by zero was wrong. He would have claimed that 7 ÷ 0 was infinity. But that doesn't really make any sense when you think about it. Even an infinite amount of nothing is still nothing — how could it ever become 7?

And that is the key to this dilemma. You need to recognize that divide and multiply operators are really the same thing. Dividing 7 by 2 makes 3.5, because when you multiply 3.5 by 2 you get 7. So when you ask what 7 divided by 0 is, you're really asking what do you multiply by zero to get 7?

The answer is there is no answer! There is no number that you can multiply by zero and get the result 7. So the answer to any problem where a number is divided by zero, is *undefined*. It doesn't make sense; it doesn't follow the rules. Generally, it's to be avoided at all costs. (It's amazing how many computer programs fail and cause the computer to crash because they accidentally told the computer to calculate something divided by zero. There's even a special computer error called the divide-by-zero error.)

The terrifying thing is that this idea (that dividing a number by zero produces an undefined result) is so counterintuitive that even some textbooks being published today claim the answer is infinity. Bhaskara's 900-year-old argument seems to lodge in the minds of people better than the true answer does. But you know better — and make sure you tell them if you see it written down incorrectly anywhere!

So what about the opposite? What about 0 ÷ 7? Well, using the same idea as we just saw, asking what number results from 0 ÷ 7 is the same as asking what number should we multiply by 7 to get zero? Put this way, the answer is obvious — to get zero we multiply by zero. So the answer to zero divided by anything must always be zero. To be fair to Brahmagupta, he did guess this one correctly.

Having sorted that out, the hardest problem of all remains. What is zero divided by zero? This is the problem that mathematicians have got wrong for centuries. We saw earlier that if you divide a number by zero the result is undefined, and we know that zero is a number, so perhaps the result of zero divided by zero is also undefined? Well, not quite.

Imagine we decided to figure out what $0 \div 0$ was by starting with bigger numbers and slowly making them smaller and smaller, making them ever-closer to zero. So we could start with 128 divided by 128, then 64 divided by 64, then 32 divided by 32, and so on. The answer seems to be 1. As this sequence gets smaller and smaller, approaching nothing over nothing, because we're dividing the same number by itself, it approaches (or "tends to" as we usually say in maths) the value of 1. But imagine we slightly change the sequence. Let's make the first value seven times bigger, and then again slowly shrink both, until we again approach nothing over nothing. Now the sequence tends toward the value of 7!

$$\frac{128}{128}, \frac{64}{64}, \frac{32}{32}, \frac{16}{16}, \cdots, \frac{0}{0} \rightarrow 1$$

$$\frac{7 \times 128}{128}, \frac{7 \times 64}{64}, \frac{7 \times 32}{32}, \frac{7 \times 16}{16}, \cdots, \frac{7 \times 0}{0} \rightarrow 7$$

Using this argument, it's simple to show that zero divided by zero could be anything at all! So the answer is not undefined (i.e., it doesn't make sense). The answer is *indeterminate* — meaning it could be any number. Divide nothing by nothing and you might get anything. Who would have thought it? No wonder Brahmagupta got the answer wrong!

Over a thousand years after Brahmagupta had written his work, a French mathematician named Guillaume de l'Hôpital received the credit for this

Above: Guillaume de l'Hôpital, who wrote the first textbook on calculus consisting of the lectures of his teacher Johann Bernoulli.

idea of shrinking series to zero over zero and it is known today as l'Hôpital's Rule. l'Hôpital was born in Paris in 1661 and began his working life as a cavalry officer, but resigned because of nearsightedness (or more likely because he was wealthy and preferred to do something else). He decided to study mathematics, and paid Swiss mathematician Johann Bernoulli a significant salary to receive private tuition, sometimes in l'Hôpital's country house at Oucques. When l'Hôpital published his book containing the now-famous rule, Bernoulli was extremely upset, for the book mainly seemed to consist of the lectures he'd given l'Hôpital. The only acknowledgment in the book for Bernoulli was rather patronizing and a little dismissive.

Above: The Swiss mathematician, Johann Bernoulli.

And then I am obliged to the gentlemen Bernoulli for their many bright ideas; particularly to the younger Mr Bernoulli who is now a professor in Groningen.

In fact, Bernoulli was so upset that after l'Hôpital died in 1704, he claimed that he was the true author of the book. Few believed him until proofs were discovered in 1922, long after both had died.

Despite the intrigue and politics, mathematics has not renamed the rule "Bernoulli's Rule," so the injustice (if that is what it was) behind the solution to $0 \div 0$ continues to this day. But as fate would have it, the name of Bernoulli is much better known in mathematics than l'Hôpital because Daniel Bernoulli, the son of Johann Bernoulli, created the "Bernoulli Principle." This is the equation that describes how, for example, a ping-pong ball will hover in a stable updraft of air. (Today there are even hugely scaled-up versions of this, where a person can hover like a stationary skydiver in the updraft created by a giant fan.) But there was another twist to the tale. Perhaps incensed by not getting sufficient credit from l'Hôpital, Johann decided he would try to take the credit for his own son's work. Johann published a book that was based on the work of his son, then changed the date to make it appear as though his book had been published before his son's. Luckily no one was fooled by this disgraceful behavior. Johann had similar feuds with his own brother and with other colleagues and students, and even tried to prove Newton's work wrong. It is perhaps fitting then, that l'Hôpital retains the credit for l'Hôpital's Rule.

The Year Zero?

Zeros don't just cause mathematicians difficulties, they've also caused quite a few problems for everyone else. For example, our calendar was first proposed by a medical doctor named Aloysius Lilius, from Calabria in southern Italy. His brother presented the idea to the pope after Lilius had died, and the Gregorian calendar was adopted in 1582. (The Lilius crater on the Moon is also named after him.) But in 1582, zero was not a commonly used number for counting. So there is no year zero in the Gregorian system — the calendar moves from 1 B.C. to 1 A.D. without anything in between.

Like all counting systems, the calendar uses ordinal numbers. These are the numbers we use to represent sequences, and so our Gregorian calendar numbers the intervals in time in much the same way that a ruler measures intervals in space. The year 1 A.D. represents the interval of time from 0 to 1 year A.D. Cardinal numbers, on the other hand, represent quantities or values independently of sequence or order. Because zero is a relatively new invention for humans, we normally use zero purely as a cardinal number, so we define quantities with it, but don't count with it.

How could we count with zero? Well, in computer science, it is actually what we always do, for zero is used a lot by computers (as we'll

Below: Pope Gregory XIII presiding over the commission for the reform of the Julian calendar. The Gregorian calendar was eventually adopted in 1582.

Below right: A brass perpetual calendar used for determining the dates of Easter in the Julian and Gregorian calendars. The Julian calendar was established in 46 B.C. by Julius Caesar. The Gregorian calendar was instituted in 1582 by Pope Gregory XIII to reform the Julian calendar.

Right: The blade of a German calendar sword (c.1686). The blade is etched with a perpetual Gregorian calendar based on the year 1686, and illustrated with signs of the zodiac.

see in the next chapters). When counting to 10 using a computer, we always start at 0 and end at 9. This method may seem strange but our calendar would be a bit more sensible if it started at zero. A.D. stands for *Anno Domini* which means "In the year of Our Lord," but by starting at 1, the calendar is incorrectly celebrating the first birthday of Christ on the day he was born. On 2 A.D., Christ was 1 year old. On 3 A.D. he was 2. (In fact the calendar is probably much more inaccurate than this, for according to Matthew 2, King Herod was alive when Jesus was born, and historical records show that Herod died in 4 B.C. according to our calendar.) So our calendar is in a bit of a mess. Because we had no zero, the beginning of the second century was actually 101 A.D. The recent millennium celebrations were all a year out — the year 2001 A.D. was really 2,000 years since Christ's (perceived) birth. Perhaps we should start learning from our computers and start counting at zero.

Unfortunately, computers don't always do so well with dates. The so-called Millennium Bug, which occurred when the date changed from

Above: Botticelli's Adoration of The Magi *(1475). The year of the birth of Christ is somewhat confused by the*

Gregorian calendar, counting year one as the year of Christ's birth rather than the year he turned 1 year old.

1999 to 2000, was caused by programmers using just two numbers to represent the year in their software. So when 99 became 00, the computers thought it was 1900 rather than 2000. In the late 1990s this caused a flurry of "Millennium Bug programmers" to update all the software, just in case some critical program that controlled electricity power stations (or electricity billing systems) got confused and decided that it would be best not to work in 1900. In the end, those two zeros added a lot of zeros to many happy computer programmers' salaries, but the Millennium Bug didn't really cause any major problems.

Numbers don't just come in whole ones, twos and threes. Before we reach those giddy heights, there is a world of small numbers that live between zero and one. We've known these must exist for centuries; all anyone has to do is take one apple and cut it in two in order to see the problem. What do you call the two equal pieces? What strange less-than-one numbers should we use? How do we write them down, speak about them or think about them?

SMALL IS BEAUTIFUL

CHAPTER 0.000000001

Today we call these types of numbers fractions, but it has taken several thousand years and many philosophers and mathematicians to understand what they should look like and why they should do what they do.

Rational Numbers

Pythagoras was perhaps the first professional explorer of the world of numbers, but he wasn't very good at fractions. He was one of the world's first mathematicians, living from 569 B.C. to around 475 B.C. (at exactly the same time that Siddhartha Gautama, who became Buddha, lived). This was long before advanced concepts such as zero had been invented, and even before the idea of division was understood. Pythagoras was born on the Greek island of Samos, and led an eventful life during which he traveled to Egypt where he was influenced greatly by their philosophers and customs, was taken as a prisoner of war to Babylon where he learned mathematics, music and sciences, before returning home to Samos, and then moving to Croton, in Southern Italy. In Croton he started a philosophical and religious sect, which soon had many followers, both men and women. Inner circle members of the sect were called *mathematikoi*, and were taught by Pythagoras to follow his strict rules. They had to give up their possessions, become vegetarians and follow his beliefs:

Above: Greek philosopher Pythagoras (shown above in a woodcutting) was also one of the world's first mathematicians.

1   That at its deepest level, reality is mathematical in nature
2   That philosophy can be used for spiritual purification
3   That the soul can rise to union with the divine
4   That certain symbols have a mystical significance
5   That all brothers of the order should observe strict loyalty and secrecy

To Pythagoras the idea of mathematics, philosophy and religion were all entwined. Indeed, an oft-quoted saying by Pythagoras is "Everything is number," which may derive from

Above: Detail of mural showing Egyptians recording the harvest. The Egyptians were innovators in creating methods of writing fractions.

Aristotle's writing a hundred years later, when he said, "The Pythagorean ... having been brought up in the study of mathematics, thought that things are numbers ... and that the whole cosmos is a scale and a number."

All of the work attributed to Pythagoras came from the Pythagorian Society, but may not necessarily have been by him. We don't know, for there are no written records from that time. It is perhaps ironic that the most well-known math theorem, Pythagoras' theorem, may have been proved by one of his followers and not him. (In fact the first statement of the theorem – that the sum of the square of the two sides of a right-angle triangle equals the square of the remaining side – was found on a Babylonian tablet, which dates from 1900–1600 B.C. – a thousand years before Pythagoras lived. Nevertheless, he or his followers may have been the first to actually prove it was

always true.) We do know that this group of mathematicians and philosophers studied geometry and believed (at least initially) that all numbers were rational. That is, every number could be expressed as either a whole number or as a ratio using two whole numbers (integers). So a quarter could be expressed using the integers one and four, perhaps like this 1:4. (The shock of discovering that irrational numbers existed was to come later, as we'll see in Chapter √2)

Pythagoras didn't quite use fractions as we know them today, but he and his followers did think long and hard about sub-multiples of numbers (or factors – the smaller numbers you multiply together to get larger ones) and ratios. In fact, Pythagoras made one of the first mathematical studies of music, discovering that when the lengths of several vibrating strings form ratios of whole numbers to each other, they created harmonious tones. Perhaps this helped his playing, for Pythagoras was also an accomplished lyre player, and would play to ill people to make them feel better.

Fractions were so important when trading (I'll trade you a quarter of a pig for a third of a sack of apples) that the Babylonians and Romans soon had symbols and words to define fractions, and the Egyptians developed a well-established method for writing fractions. The fractional notation that we know today (1 over a 3 to mean one-third) became a more common usage by the time our friend Brahmagupta wrote his book in 628 A.D. By then, a number was placed over another without the horizontal line, to indicate a fraction. The line did not appear for another 600 years. Fibonacci was perhaps the first European to write fractions in the way we do today, but we'll be seeing much more of him later.

An Important Period

While we got used to the idea that numbers might not always define a whole thing – a number might be a fraction of a whole – the idea of the decimal point took longer to emerge. This was despite the widespread use of the abacus, which nearly expressed fractions in the way we know today.

The Romans used pebbles to perform their arithmetic. They used counting boards that had grooves cut into them to hold the pebbles, and each groove corresponded to a numerical value (1,000, 100, 10, 1, ½, ⅓, ¼). When a Roman spoke of doing arithmetic, he would use the words *calculus ponere* ("to place pebbles") – which is where the words calculate and calculator come from. Their method of defining numbers with fractions is surprisingly similar to the way in which numbers with fractions are represented using binary in computers today. For example, one simple method is to use a 1 or a 0 in a specific position to define the parts of a number, so

0	1	1	1	0	1	1	0
8	4	2	1	½	¼	⅛	1/16

represents $4+2+1+\frac{1}{4}+\frac{1}{8} = 7\frac{3}{8}$. Replace the 1s with pebbles and you have something that looks just like the Roman abacus.

Although this is very close to the idea of adding a decimal point and writing fractions as decimal fractions, it took nearly a thousand years

Below: Egyptian mathematical papyrus, dating from c.1550 B.C.

Above: Wooden Chinese abacuses like this one have been used for over 700 years.

A decimal fraction is always the same: 3.375, which makes doing sums much simpler.

The abacus was quickly adapted to exploit this innovation. It reached its pinnacle of development when Roman pebbles were turned into beads and threaded onto bars. The Chinese abacus is perhaps the most successful example of this design. Its familiar form (still used for teaching in some Chinese schools today) was first introduced around 700 years ago. Each rod represents a different value: one for units, tens, hundreds, thousands and so on. The rightmost rods represent tenths and hundredths – the first two decimal places. This meant that as well as being able to represent very large numbers (on a typical abacus with 10 rods, numbers up to 99,999,999.99 could be calculated) the Chinese abacus could also represent numbers as small as 0.01. The speed at which a trained abacus user can calculate sums is remarkable, and proficient users are even able to visualize the movement of beads in their heads in order to achieve astonishing feats of mental arithmetic.

Thinking Small

Once we could write tiny numbers as fractions and as decimal fractions, a whole new world was opened up to us. We could now think about things so small that we cannot even see them. And more importantly, we could describe exactly how small they are.

We now know that much of the world is invisible to us because it is too small for the eye

before a Syrian mathematician figured it out. Abu'l Hasan Ahmad ibn Ibrahim Al-Uqlidisi was born at around 920 A.D., possibly in Damascus. He wrote the earliest text we know of that describes how to write 7.375 instead of $7\frac{3}{8}$. The advantages of this system were as obvious as moving to a positional system that used zero was – now highly precise arithmetic could be calculated by lining up all the numbers on each side of the decimal point. It also made numbers less ambiguous. Rather confusingly, the same fractions can be written using different numbers, so $7\frac{3}{8}$ is the same as $\frac{59}{8}$, which is the same as $\frac{118}{16}$, which is the same as $\frac{177}{24}$, and so on.

to see. We know that we're made from trillions of cells, each about 100,000 times smaller than our height, that's between about 7 and 30 micrometers, or 0.000007 and 0.00003 meters in size. We also know that the viruses that infect our cells are a hundred times smaller than that, ranging from 20 nanometers (polio) to 300 nanometers (smallpox) or 0.00000002 to 0.0000003 meters. Viruses are the simplest form of life, really not much more than complicated molecules, and molecules are made from atoms. A hydrogen atom is very small; a thousand times smaller than a virus – only 0.5 Ångstrom across, or 0.00000000005 meters in diameter. Atoms are made from even smaller things called protons, neutrons and electrons. A proton is several thousand times smaller than an atom at around 10 femtometers, or 0.00000000000001 meters. And protons are made from quarks, which are a thousand times smaller still – about 10 attometers or 0.00000000000000001 meters. But the winner – if it exists at all – would be the strings hypothesized by physicists in the exotic "String Theory" to exist right at the bottom. One string would be about 0.0000000000000000000000000000000001 meters across.

Nanotechnology is the technology devoted to nano-sized devices. (The title of this chapter is

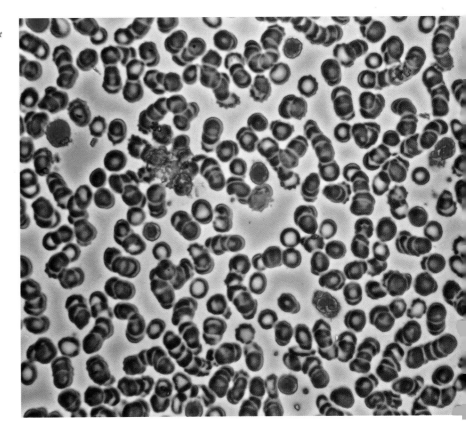

Right: We are all made from trillions of cells; the viruses that infect our cells are a hundred times smaller than the cells they infect.

1 nanometer, as well as being the decimal fraction of 1/1,000,000,000.) DNA, the molecule that holds together and makes up all our genes, is 2 nanometers (nm) in diameter (although if it was stretched out in a line, it would be 6 feet (1.8 m) long – it's a long and very coiled-up molecule within each of our cells). The only function of a gene is to produce a protein (a complicated molecule made from amino acids) and those proteins are the clever chemicals that tell cells what to do and where to go. Proteins are about 3 to 10 nm in size. We're still learning how to manipulate things at nanoscales, but a variety of tiny devices now exist. In 2003 scientists at the University of California, Berkeley, created the smallest electric motor, less than 500 nm in size. Silicon chips are also rapidly shrinking – the smallest transistor to date has been just 50 nm

across. Scientists at the Massachusetts Institute of Technology have even managed to attach a nanosized radio antenna onto a gene and use a radio signal to control its expression. Radio controlled biology, here we come!

So fractional numbers have transformed our ability to think small and understand the dimensions of things like atoms. But there is a twist in this tale. Two and a half thousand years ago, at the same time that Pythagoras was learning about numbers, the young Buddha had already mastered a seemingly impossible trick with tiny numbers.

The record of this story comes from a "biography" of Buddha, the *Lalitavistara Sutra* ("the development of games," written in verse and prose). Siddhartha Gautama, or Bodhisattva (as he was called before he became Buddha) was born around 565 B.C. in the city of Kapilavastu, northern India, today known as Nepal. In the *Lalitavistara Sutra* a competition is described between Bodhisattva and a mathematician called Arjuna who is seemingly very impressed with the young Bodhisattva's knowledge. Arjuna asks how one describes the smallest possible particle or "first atom." Bodhisattva explained how the sizes of different tiny things were related in multiples of 7. The answer was long, but essentially said:

Below: Colored scanning electron micrograph of the drive gear (orange) in a micromotor.

This gear is smaller in diameter than a human hair and 100 times thinner than a sheet of paper.

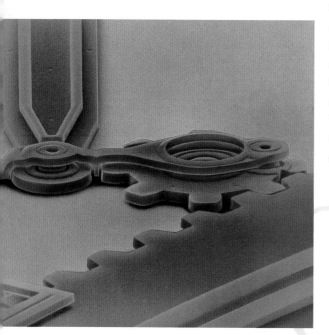

There are 7 "first atoms" (*paramanu raja*) in 1 minute particle of dust (*renu*),

7 of the latter in 1 tiny speck of dust (*truti*),

7 of those in 1 speck of dusk carried by the wind (*vayayana raja*),

7 of those in 1 speck of dust stirred up by a hare (*shasha raja*),

7 of those in 1 speck of dust stirred up by a ram (*edaka raja*),

7 of those in 1 speck of dust stirred up by a cow (*go raja*),

7 of those in 1 poppy seed (*liksha raja*),

7 poppy seeds in 1 mustard seed (*sarshapa*),

7 mustard seeds in 1 grain of barley (*yava*),

and 7 grains of barley in 1 phalanx of a finger (*anguli parva*).

If we say that 1 phalanx (long bone) of a finger is about 4 cm, we can work out how small a measurement the Buddha was talking about for the "first atoms."

$$0.04 \div 7 \div 7 \div 7 \div 7 \div 7 \div 7 \div 7 \div 7 \div 7 = 0.0000000001416 \text{ or } 1.416 \times 10^{-10}$$

That's 141.6 picometers or 1.416 Ångstrom, which rather remarkably happens to be pretty much the size of a carbon atom. Not bad for 2,500 years ago, before anyone even knew atoms existed!

Above: Statue of the Buddha, who explained how the sizes of tiny things were related in multiples of seven.

The first number you learned and probably the first you spoke: "one." One has always been a number full of meaning. Since we realized that four quarters magically turn into a whole one, the number 1 has meant single, wholeness, unity and togetherness.

ALL IS ONE

CHAPTER 1

Above: "Four quarters turn into one," giving the number one a single wholeness ... as well as a variety of superstitious connotations, both good and bad.

There are many superstitions about the number one, and some even have a little sense behind them. It is said that "if you break one egg, you will break a leg" — a good rhyme to teach clumsy children to be careful with eggs. Or "it is unlucky to walk around the house in one slipper" — which is clearly true when you stub your toe. Some say "only keep money in one pocket or you will lose it" — which again, seems quite sensible. But then there are the less-pleasant superstitions, whose origins we can only wonder at. For example, apparently "people with one hand are psychic." Did some poor one-handed person get sick of being teased and warn the bullies away? Or there's the rather nasty, "a one-eyed person is a witch" — which sounds more like an insult than a superstition. Some superstitions about the number one are related to animals. For example, "seeing one magpie bodes a death in your future" or "seeing one white horse brings bad luck." And there are even omens related to dates. Apparently, "if you wash your hair on the first day of the month you will have a short life" —

something that thankfully does not seem to be true, given how frequently we all wash our hair in modern times. And if you're thinking of marriage, then apparently, "it is unlucky to get married August 1 or January 1."

One is a number that can cause more than trepidation. In many religions, the number one is associated with the unity of God. It was believed that if you dreamed about the number one, you had received a direct message from God. But unfortunately members of each religion believed that there is only one God and that the exclusiveness of one belonged to them, the true believers. Had they all believed there were, say five gods, then there might have been a little more room for compromise. But the number one has no compromise. It is a single, unequivocal, exclusive oneness. That oneness has led to intolerance and centuries of bitter, bloody battles.

Thankfully the number one has nicer connotations as well. The mysterious Philosopher's Stone — the unknown catalyst that was thought to transform base metals

into gold and be the elixir of everlasting life — was said to be "one in essence" according to the 17th-century scholar of alchemy William Gratacolle. (However, William also found about a hundred other names for the Philosopher's Stone, ranging from "eyes of fishes" to the rather bizarre "belly of a man in the mist," so it is unlikely that being "one in essence" was terribly helpful in the centuries-long search for the stone.)

And of course, number one is often used to mean "the best." In China, Feng Shui can be improved with the right numbers. The number one is the first of the yang numbers and is strongly linked to growth and prosperity. But in the West, strangely, we don't always use the word "one" in this way. For example, when talking about playing cards, the one of hearts is always called the ace of hearts. It is thought that the word ace actually derives from Roman money, for the lowest value Roman coin was called an "as" and also meant "a whole" or "one unit." The meaning of the name "Ace" (which some people have as a surname) means "number one" to this day.

Natural Numbers

One of the reasons why people over the centuries have believed that the number one has deep mystical significance is because of mathematicians. Numbers aren't just small, big or rational as we saw in the last chapter. Some of them are natural, some are perfect, others are amicable and some are prime. These numbers are special because they have rare properties. Since only some numbers have these properties, strange new patterns of numbers emerge: patterns within patterns. Some of these patterns are so old that their discovery predates Pythagoras, but we do

Above: Diagram of alchemical process for production of the Philisopher's Stone, which was said to transform metals into gold and be "one in essence."

Right: Historical artwork taken from A Very Brief Tract Concerning the Philosophical Stone, *Frankfurt, 1678. This shows many of the symbols associated with alchemy. Alchemy, the pseudo-scientific predecessor to chemistry, is best known for its practitioners' hunt for the Philosopher's Stone, which would impart eternal life and the ability to turn base metals into gold.*

know that the Pythagorian *mathematikoi* were deeply fascinated by special types of numbers. At the same time that the young Buddha was learning many of the truths that would become Buddhism, the Pythagoreans were studying the universe through numbers. They believed that numbers were fundamental to the universe — that in a very real sense, everything is built from numbers. So by discovering and analyzing patterns of numbers, they believed that they might be decoding some deep meanings that would help explain how and why the universe was the way it was.

Every child is first taught the natural numbers — these are the numbers we count with on our fingers: 1, 2, 3, 4 ... We call the set of non-negative, whole numbers *natural* because they are very natural — they're the most common,

most frequently seen of all numbers. Natural numbers are the pigeons of the number world — you see them everywhere you look. Just take a look around you now and open your eyes to the natural numbers surrounding you. They unobtrusively sit on almost every product made by us. And if you see none written down — look closer. How many trees do you see? How many clouds, or windows or people? Those are natural numbers growing in your brain as you count.

Traditionally, we have considered the natural numbers to begin with one. This was not always the case — the Ancient Greeks thought one was a "unit" and the natural numbers began with two, as it represents a "multiplicity." But as studies of the natural numbers continued, most mathematicians soon concluded that one was the first and therefore most natural of numbers. After all, all other natural numbers can be made from adding the right number of ones together. There can be no more natural number than the number one. They also thought in this way because (as we saw earlier), counting with zero is not natural to us. Today there is still some disagreement — some believe zero is natural, others don't. In the end, it doesn't matter too much, as long as we're consistent.

One advantage of starting the natural numbers with zero is we can define the meaning of operators on the natural numbers using it. So we can actually define what addition means, for all natural numbers. You might ask why we would need to do that — surely addition is obvious. But is it really? How do you know that adding 1 to a natural number will

increase its value by 1? What if there's a weird natural number out there that doesn't obey this rule? In mathematics, nothing is taken for granted. Everything is defined and nailed down tight, which is why when we prove something mathematically, we know for sure it is true.

As well as proving what addition does, one plays an important role in multiplication. It is called the multiplicative identity, because it has no effect. One multiplied by any number gives the same

Defining addition

How do you define addition? It's not as though you can list every possible sum and give the answer. One way to do it is to provide some unbreakable axioms:

$a + 0 = a$
$a + S(b) = S(a + b)$
$S(0) = 1$

where S is the successor function (which means "count one more" and simply gives the next natural number) and a and b may be replaced with any natural number.

The successor function uses the notion of counting, and using these axioms we can use mathematics to say, "if you can count, you can add," by defining addition in terms of the successor function.

So if we want to know what $a + 1$ actually means for any value of a, we can use the

axioms and replace 1 with S(0), for we know the successor of 0 is 1:

$a + 1 = a + S(0)$

then we can use another axiom with b taking the value of 0 and rewrite the same expression:

$a + S(0) = S(a + 0)$

and then finally use the remaining axiom to replace $a + 0$ with a:

$S(a + 0) = S(a)$

So we have proved that $a + 1$ is exactly the same as $S(a)$ for any value of a.

Every operator (e.g., subtract, multiply, divide) is defined equally carefully in mathematics, so we always know that they will do what we want them to.

number, which is an important axiom or truth in mathematics. (Zero is the additive identity as we saw in the first axiom of the addition proof.) Mathematics is built upon these axioms so we like to keep them simple, and clearly true.

Perfect Numbers

Some of the first patterns of natural numbers spotted and studied by the Pythagoreans were perfect numbers. These are natural numbers with a bizarre property that they can be formed by adding up all the smaller numbers that make up their divisors. The number 6 is the first perfect number, because it may be divided by 1, 2 and 3, and those same numbers add up to 6. Not many numbers are perfect. The number 8 certainly isn't, for 1, 2 and 4 only add up to 7. Nor is the number 9, for 1 and 3 only add to 4. In fact, perfect numbers are so rare that there are only 4 of them in the first 10 million natural numbers. Those four are:

$$6 = 1 + 2 + 3,$$

$$28 = 1 + 2 + 4 + 7 + 14,$$

$$496 = 1 + 2 + 4 + 8 + 16 + 31 + 62 + 124 + 248$$

$$8,128 = 1 + 2 + 4 + 8 + 16 + 32 + 64 + 127 + 254 + 508 + 1016 + 2,032 + 4,064$$

By the time we reach the 20th perfect number, its size has become unthinkably huge. The 20th perfect number has an astonishing 5,834 digits (it's so big it would fill this entire page).

The first four perfect numbers have been known for over 2,000 years. Their significance has been debated since their discovery. When mathematics, philosophy and religion were all aspects of the same thing, it was perhaps

Above: Detail showing
St. Augustine from a mosaic
in San Paolo Fuori Lemura.

natural for people to believe that perfect numbers must be favored by God. For example, St. Augustine (354–430) wrote in his famous text *The City of God*:

Six is a number perfect in itself, and not because God created all things in six days; rather, the converse is true. God created all things in six days because the number is perfect ...

Likewise, 28 was thought to be chosen by God as the perfect number of days to take the Moon to orbit the Earth (although today we know that the number is closer to 27.322 days).

But perhaps René Descartes summarized perfect numbers best when he said:

Perfect numbers like perfect men are very rare.

Below: Illustration showing the Moon orbiting the Earth

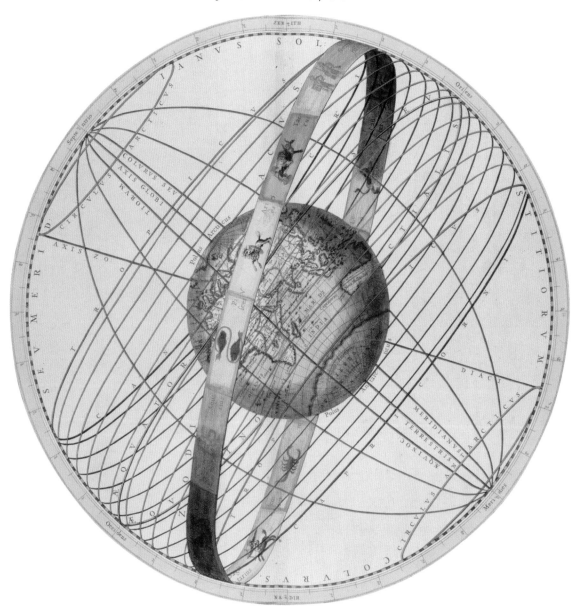

Amicable Numbers

While very few numbers are perfect, some friendly pairs of numbers occur a little more frequently. Amicable numbers as they are called, are numbers whose divisors add up to their "friend." The two best known examples of amicable numbers are 220 and 284:

220 divisors are:

1 + 2 + 4 + 5 + 10 + 11 + 20 + 22 + 44 + 55 + 110 = 284

284 divisors are: 1 + 2 + 4 + 71 + 142 = 220

Amicable numbers have been known as long as perfect numbers, but their significance has been regarded as being quite different. While perfect numbers were considered to be like mystical columns that held up the universe, amicable numbers were treated as perfect partners, indicating two things meant to be together. Not surprisingly, it is said that 2,000 years ago, couples in love would exchange talismans and lockets with the numbers 284 and 220 on them, and apparently many marriages have happened between partners who each had perfect numbers for birthdays, horoscopes, heights (or anything else they could think of). According to some, when Pythagoras was asked what a friend was, he replied that a friend was one who is the other I, such as 220 and 284.

There was once even an experiment performed by an Arab in the 11th century who decided to test the resulting erotic effect of having one person eat something labeled 220

Above: Illustration showing Pierre de Fermat, the "Founder of modern theory of numbers."

and the other eating something labelled 284 at the same time. Sadly he did not report how the experiment worked out.

Despite the huge interest in amicable numbers, for centuries only 220 and 284 were known. It is often written that Pierre de Fermat (the same Fermat who caused the controversy with his last theorem, which we'll return to later) discovered the next pair: 17,296 and 18,416. But actually, he did not — an Arab called Ibn al-Banna beat him to it by several hundred years.

Descartes (yes, the same fellow who said "I think, therefore I am") found the third pair of amicable numbers: 9,363,584 and 9,437,056 (although some claim that they may also have been known before this). By 1747 the well-known mathematician Euler had discovered over 30 amicable numbers (although embarrassingly he actually got some wrong). But perhaps the most startling finding was made in 1866, when a 16-year-old Italian boy called Paganini discovered the friendly pair: 1,184 and 1,210. These formed the second lowest pair and had been completely overlooked despite some of the best mathematicians of the previous 2,000 years studying the problem.

Prime Numbers

Perfect and amicable numbers are made by worrying about their divisors — the smaller natural numbers that you must multiply together to make them up. As we've seen, 6 has divisors of 1, 2 and 3, as those are the numbers that "fit inside" 6 perfectly. But what if there are numbers that no other numbers "fit inside"? We have to exclude 1, for 1 clearly fits inside all natural numbers (and we proved that earlier). When we consider which numbers have no divisors except for 1 or themselves, we are thinking about prime numbers.

Prime numbers are truly special. They can't be perfect or amicable, for they may only be divided by 1 (or themselves) and produce a natural number. Despite being so special, there are an awful lot of them. The first 10 prime numbers are 2, 3, 5, 7, 11, 13, 17, 19, 23 and 29. But they just keep on going. They're not nearly as hard to find as perfect or amicable numbers, and in fact are strongly related to those other special numbers.

In Stephen King's novel, *The Dark Tower III: The Waste Lands*, the heroes of the story Roland, Eddie, Jake, Susannah (Detta) and Oy are desperate to leave a decaying city on the only remaining train. That train was intelligent, and somewhat malevolent, and so it demanded they answer a riddle before they could board. The riddle was "You'll have to prime the pump to get me going and my pump primes backward." It was Detta who worked out that the train wanted the prime numbers entered on its diamond-shaped keypad, backward. So she set about working out what they were:

Prime numbah be like me — ornery andforspecial. It gotta be a numbah you make by addin' two other numbahs, and it don't never divide 'ceptin' by one and its ownself. One is prime just 'cause it is. Two is prime, 'cause you can divide it by one an' two, but it's the only even number that's prime. You c'n take out all the res' dat's even.

"I'm lost," Eddie said.
"That's 'cause you just a stupid white boy," Detta said, but not unkindly. She looked closely at the diamond shape a moment longer, then quickly began to touch the tip of the charcoal to all the even numbered pads, leaving black smudges on them.

Above: Portrait showing side profile of Eratosthenes.

"Three's prime, but no product you git by multiplyin' three can be prime," she said.

Susannah began using her charcoal to touch the multiples of three which were left now that the even numbers had been eliminated: nine, fifteen, twenty-one, and so on.

Same with five and seven, she murmured. "You just have to mark the odd ones like twenty-five that haven't been crossed out already.
"There," she said tiredly. "What's left in the net are all the prime numbers between one and one hundred. I'm pretty sure that's the combination that opens the gate."

And it was, allowing the group of travelers to continue their difficult journey.

In the story, Susannah used the "Sieve of Eratosthenes" to figure out the prime numbers. But Eratosthenes was not fictional. He was a scholar born in Cyrene, North Africa (now Libya), in 276 B.C. Eratosthenes was a very successful scholar, who figured out the circumference of the Earth with surprising accuracy, as well as the tilt of the Earth's axis, a calendar complete with leap years and even

catalogued nearly 700 stars. His "sieve" was a method for finding primes, in just the way described in the Stephen King novel. Like Susannah in the story, Eratosthenes also may have believed that 1 is the first prime number. Indeed, for many centuries 1 was thought to be the first prime number. After all, you can divide 1 by 1, and by itself (also 1) and it fits exactly. But in recent years we have chosen to omit 1 from the list of primes, mainly because of someone called Euclid.

Although the Pythagoreans and other mathematicians of that time were fascinated by prime numbers and no doubt attached deep mystical significance to them, it was a mathematician called Euclid who made the first breakthroughs. Euclid was born in around 325 B.C. and spent his adult life in Alexandria, Egypt. Little is known about his life except his works on mathematics (and some argue that Euclid's achievements may have been the work of several mathematicians rather than that of a single man). But some evidence does suggest that not only did Euclid the man exist, but he may have also had a good sense of humor. According to the Greek writer Stobaeus:

> ... someone who had begun to learn geometry with Euclid, when he had learnt the first theorem, asked Euclid 'What shall I get by learning these things?' Euclid called his slave and said 'Give him threepence since he must make gain out of what he learns.'

It seems that students have asked the same questions of their mathematics teachers since math was invented. Perhaps Euclid's answer was better than most, though.

Euclid's most well-known work was called *Elements*, and this remarkable 13-book epic laid the foundations of modern mathematics. It is claimed by some that next to the Bible, *Elements* may be the most translated, published and studied

Below: Arabic translation of Euclid's Elements.

Above: Illustration of the astronomer and geographer Ptolemy and the mathematician and physicist Euclid.

of all the books produced in the Western world (and of course it is much older than the Bible). It is said to be the greatest mathematical textbook of all time. Many of the books deal with geometry, defining important concepts and properties of triangles, rectangles, circles, proportion, plane geometry and three-dimensional geometry. These concepts remain valid to this day, with Euclidian geometry being our mainstay for architecture and design throughout the modern word (we'll return to this topic in a later chapter). In books 7–9, he focuses on number theory. In one book he is even thorough enough to give a definition of the number 1:

A unit is that by virtue of which each of the things that exist are called one.

Euclid's thoroughness was legendary. By defining all the truths he could think of, he was able to build many proofs about numbers and shapes that we have used ever since. One of his most wide-reaching proofs was about prime numbers. Euclid managed to prove the most important thing about natural numbers and prime numbers ever discovered. It is so important that today we call it "The Fundamental Theorem of Arithmetic." That theorem says:

Every natural number greater than one is either a prime number or it can be written uniquely as a product of primes.

To see what this means, think of a number. Now work out which natural numbers you need

Above: Original reproduction of Euclid's Elements.

on hope, we'd still be calling it a theory. No, Euclid proved his idea was true, and he did it using one of the earliest known examples of a proof by contradiction. This form of proof is based on the idea that if I claim something is always true, then when I try to imagine a counterexample that would make it false, the result makes no sense. So, for example, we could prove that not everything is true using this type of proof:

My theory is that all beliefs are equally true and cannot be denied.

Harry believes in a flying spaghetti monster that orbits the sun.

I deny the existence of the spaghetti monster.

But according the theory, Harry's belief is true AND my belief is true, yet we believe the opposite of each other. Harry thinks he's right and I think he's wrong. We can't both be right, so the theory must be wrong.

So Euclid wanted to prove that all natural numbers greater than one are made from products of prime numbers. He tried to imagine a counter-example: some natural number that could not be made from products of primes. Perhaps, in fact, there might be more than one that could not be made from products of primes. But he only needed a single number to disprove his theory, so he imagined choosing the very smallest one. This hypothetical number must be a product of at least two other numbers: $a \times b$, and those numbers must not be prime

to multiply together in order to make that number. Euclid's theorem tells us that those factors will be prime numbers. (If you thought of a prime number to begin with then you have nothing to do.) Don't believe Euclid? Well, let's try it with 72. You can make 72 by multiplying 18 and 4. You can make 18 by multiplying 9 and 2, and you can make 4 by multiplying 2 and 2. So the smallest factors of 72 are 2 x 2 x 2 x 9. And, you guessed it, 9 and 2 are prime numbers. According to Euclid, this will work for any natural number.

Because he was a mathematician, he didn't just hope that his theorem was true. If he relied

numbers. But Euclid chose the smallest number that is not a product of primes, so a and b must be products of primes (otherwise we contradict the fact that he chose the smallest). But if they are products of primes, then the number created by multiplying them must be a product of primes, so we contradict the counterexample. That's how he did it.

He used the same trick to prove that there are infinitely many prime numbers (there is no way you cannot find one that's just a little bit bigger than the last).

Euclid also showed that prime numbers and perfect numbers (whose factors add up to the value of the number, if you recall) are highly related. He showed that if a prime number can be made by adding a sequence of numbers that double each time, then the prime number multiplied by the largest factor will be a perfect number. So we can make the prime number 7 by adding this sequence:

$$1 + 2 + 4 = 7$$

and (the sum) x (the last) = 7 x 4 = 28, which is a perfect number.

Or we can make the prime number 31 by adding this sequence:

$$1 + 2 + 4 + 8 + 16 = 31$$

Then 31 x 16 = 496, which is a perfect number.

Amazingly it took nearly 2,000 years before another mathematician called Euler was able to show that all even perfect numbers took this form. To this day, it is not known whether there are any odd perfect numbers. You're welcome to look for some!

But although we could now use some primes to find perfect numbers, because of the fundamental theorem of arithmetic, the

Below: Leonhard Euler depicted at his desk.

number 1 became an outcast. It wasn't a helpful number — in fact it got in the way of the theorem — and so, about 300 years ago everyone agreed to exclude 1 from the primes (apart from the occasional mistakes). Just to make sure it can never join the prime gang, a new rule was added. Rather unfairly, it says "a prime number must be greater than 1." Perhaps the train in *The Dark Tower* had a rusty computer for a brain, for if it had known that rule, the heroes of the story would have failed the riddle.

Secure Primes

Primes are not too difficult for a computer to calculate, but it does take some time. The Sieve of Eratosthenes is not a very good way of working out really big prime numbers, so most prime-number generators figure out new prime numbers by using smaller ones (for example, you can see if your guess is prime by checking to see if all the smaller primes are factors). Today there are some clever ways of making primes, but still there are some special prime numbers that are exceptionally hard to find. These are called strong primes. A prime number is strong if the average value of the two primes on each side of the prime is less than the value of the prime. So, 17 is the seventh prime. The sixth and eighth primes,

13 and 19, add up to 32, and half that is 16. That's less than 17, so 17 is a strong prime. There's also another tricky type of prime called a safe prime, which is made by multiplying another prime by 2 and adding 1. Figuring out whether a huge number is a prime, especially when it's safe or strong, is hard even for a powerful computer.

That's why the security systems that we use on the internet today use cryptographically strong prime numbers. These are *big* prime numbers designed to take years for our best computers to figure out factors. Prime numbers form the basis of computer encryption, used universally to encrypt computer files. The next time you buy something online, remember that the secure payment you make is secure because of the prime numbers inside.

Fractionally One

The number 1 may not be a prime number, but it's still more than enough to confuse people. For example, there is a particular dilemma about 1 that bamboozles people. Take one apple, cut it into three. You get three-thirds. So what do you get if you multiply one-third by three? One? But are you sure about that?

In fractional notation, the answer is obvious:

⅓ x 3 = 1

But what happens if we write the same thing out as decimal fractions?

0.33333333... x 3 = 0.99999999...

This leaves us with the dilemma: where did the missing bit go? Does the decimal fraction introduce an error so that we get the wrong answer, or is 0.99999999... (with a neverending number of 9s) actually another way of writing 1?

The answer is nowhere, no and yes, respectively. Although it seems a little strange, 0.99999999... is 1. It's just a rather clumsy way of writing 1. There are many ways to prove it, but perhaps one of the easiest is to use some simple algebra (see box, right. The next chapter explains where algebra came from).

It's much easier to write 1, so that's what we usually do. But just as we can write some ideas using different words (for example, "little" can be written as petite, tiny or small), so we have different ways of writing 1. We could write 0.99999999... or we could write ¹⁄₁ or ⁴³⁄₄₃ or even (10 − 5) / (26 − 21). But what we *mean* is the single, solitary natural number, 1.

Why does 0.99999999... = 1?

If we make it 10 times bigger we get:

0.99999999... x 10 = 9.99999999...

Now if we take the first number from the second:

  9.99999999...
− 0.99999999...
= 9.00000000

You can see the answer must be 9, for except for the first digit, which is 9–0, all the others are 9–9 which become 0.

Now for something new. Let's give 0.99999999... a new name. We'll call it *a*.

If we repeat the same sum we just did, but this time using a, we now have:

$10a − a = 9$

Ten of anything take one must be nine, so we can rewrite as:

$9a = 9$

and if nine of something is nine, then dividing both sides by nine tells us:

$a = 1$

Finally, let's remind ourselves what *a* was a new name for:

0.99999999... = 1

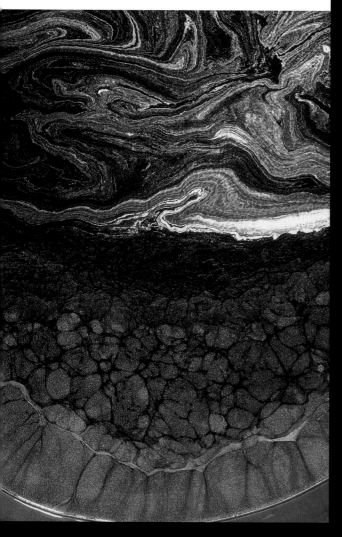

Numbers are elegant patterns embedded in the fabric of our universe. Prime numbers, perfect numbers and all the natural numbers and fractions show the richness and diversity of using numbers as a language. Everything good and just, everything that makes sense, and all the ratios of life can be described completely by these numbers. At least, that's what the Pythagoreans thought. But they were wrong.

MURDERING

CHAPTER √2

They discovered their error quite early on, but it was considered so shocking and heretical that the truth was suppressed. Ironically, the truth came from one of the great achievements of the Pythagoreans — the famous Pythagorian theorem, which states:

> The sum of the squares of the sides of a right-angle triangle is equal to the square of the remaining side.

It's a lovely Pythagorian result that shows the order and simplicity of numbers behind shape. For example, imagine we have a right-angle triangle with smaller sides 3 inches and 4 inches in length. We can use the theorem to figure out that the largest side must be of size 5 inches in length, because:

$$3 \times 3 + 4 \times 4 = 5 \times 5$$

This works for any right-angle triangle. Given the lengths of two of the sides, we can use this theorem to calculate the length of the third. Sounds perfect. But there was a very big problem. Imagine we have an ordinary square — let's say it's 1 foot by 1 foot in size, and we draw a line through the center, from one corner to another. We have made two right-angle triangles that have smaller sides all 1 foot long. What is the length of the diagonal? According to the famous theorem:

$$1 \times 1 + 1 \times 1 = a \times a$$

(a is the length we'd like to know. This is a pretty easy thing to work out (1 multiplied by 1 is 1, and 1 plus 1 is 2):

$$2 = a \times a$$

So we know that the length of the diagonal when multiplied by itself gives the value of 2. What number is that? It must be bigger than 1, as $1 \times 1 = 1$, which is much too small. It must be smaller than 2, as $2 \times 2 = 4$, which is much too big. So the answer must be a fraction of a whole number. What about $\frac{7}{5}$? That gives 1.96 when multiplied by itself. What about $\frac{707}{500}$?

IRRATIONALS

That gives 1.999396 when multiplied by itself. What about $^{7072}/_{5000}$? That gives 2.00052736.

The shocking truth is that there is no fraction that you can multiply by itself and get 2. So if there is no natural number and no rational number that, when squared, equals 2, a mysterious new type of number must exist. An unnatural number. A number that we can't write down. It has a mysterious, unknowable value. We call these numbers *irrational*.

To the Pythagoreans, the whole notion was appalling. And it got worse. If you looked at any square, you got the same result. If the squares were of side 2 m, we need a number that, when squared, equals 8. Or if the sides were 3 m, we need a number that, when squared, equals 18. None of them could be written down as fractions or integers. Just how many irrational numbers were there? To have these non-numbers appear so often must have been like a constant attack of their religious beliefs. So they did what any good religious sect does — they suppressed the truth, and pretended the "unmentionable" numbers did not exist.

But the truth came out in the end. Sometime after Pythagoras had died, the society was becoming unpopular. Resentment had built up against its secrecy and exclusiveness and the villagers drove the Pythagoreans out of Croton in riots. One prominent disciple named Hippasus decided the time was right to reveal some of the secrets — including the existence of the irrational numbers. He broke the oath of secrecy and was immediately expelled from the order. Hippasus' escape from the Pythagoreans was fleeting. He decided to become a geometry teacher, and after his expulsion went sailing. Hippasus never returned, drowning at sea. Some said the gods took vengeance on him for his betrayal. Others said he was murdered by vengeful Pythagoreans.

The Pythagoreans didn't survive much longer themselves. Although societies had spread to several other Italian cities, only a few years after the death of Pythagoras the Pythagoreans split into factions and became political. In 460 B.C. all the meeting places of the society were burned and destroyed; a surviving record tells the tale of the "house of Milo" in Croton where over 50 Pythagoreans were murdered.

With no Pythagoreans left to tell us, we'll never know what really happened to Hippasus on his boat, but we do know that the irrational numbers were too big a secret to keep. Once revealed to the world, they would never be forgotten.

Being Irrational

Today, irrational numbers are understood very well. We know that natural numbers and fractions are really more like islands of order in an endless ocean of disorder.

There are an infinite number of rational numbers (you can always add 1 to the current number). But there are *more* irrational numbers. In between every two integers or every two fractions, there are an infinite number of irrational numbers. And in between any two irrational numbers, there are an infinite number more.

So what is an irrational number, anyway? The example produced by that pesky 1 foot square was the square root of 2, written $\sqrt{2}$ (and often simply called "root two"). It's a number that cannot be written in full. We can write some of it:

1.414213562373095...

but it keeps on going forever, never forming a regular pattern. This is quite different to a rational number. For example 3/7 is written:

0.428571 428571 428571...

Above: Photograph of George Ferdinand Cantor, the German mathematician who helped to develop countability proofs.

which also goes on forever, but has a regular repeating pattern. All rational numbers are patterns. All irrational numbers are devoid of pattern. They are anti-pattern. Irrational numbers form the spaces between all the patterns.

We owe much of our knowledge on this subject to Georg Cantor, who was born in St. Petersburg, Russia, in 1845. Among Cantor's many achievements were his proofs about countability. Cantor was fascinated by infinite sets of numbers and wanted to know whether it was possible to count them or not. While it might take forever to count a set of infinite things, it is at least theoretically possible. But Cantor discovered that some things are simply not countable. The set of irrational numbers cannot be counted, for example. It's possible to understand intuitively why this is so — for to count something we need to be able to number it: 1,2,3,4,5 … If we were foolish enough to want to count integers, it's very clear that we can manage it: 1 one, 2 two, 3 three, 4 four, 5 five … But how do you count irrational numbers? We can't even write down an irrational number, so how do we know what the next one is? If we add an infinitely small amount to √2 do we get the next irrational number? But then, what about the number that falls between those two? Cantor realized (and proved) that they simply cannot be counted.

Even though it seems bizarre, some sets of infinite things are bigger than others. There are more irrational numbers than rational ones, even though there are an infinite number of both.

As well as proving things about irrational numbers and infinite sets, Cantor also had an obsession with Shakespeare. He believed that Francis Bacon was the true author of Shakespeare's plays, and spent a significant amount of time in his final years researching, publishing pamphlets and lecturing on this topic. In 1911 Cantor was invited as a distinguished foreign mathematician to the 500th anniversary of the founding of the University of St. Andrews, Scotland. Unfortunately he took the occasion to speak about Francis Bacon and Shakespeare rather than mathematics, and according to Cantor's biographer, "During the visit he began to behave eccentrically, talking at great length on the Bacon-Shakespeare question."

Cantor also sadly suffered from depression throughout much of his life and spent some time in sanatoria. He died in 1917 at the age of 72 in a sanatorium, despite writing regular letters to his wife asking to be released. Nevertheless, despite his sad end, fellow mathematician Hilbert praised Cantor's work as:

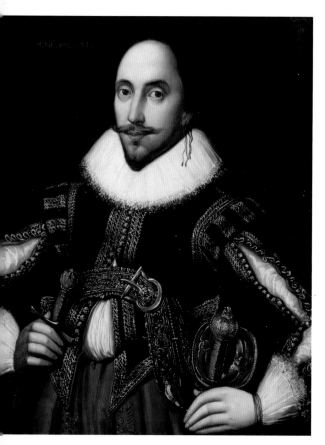

Below: Portrait of William Shakespeare, with whom Cantor was obsessed.

Above: The Egyptian Nile flooded every year, erasing important land boundaries. The division of land each year thus became known as "geometry."

... the finest product of mathematical genius and one of the supreme achievements of purely intellectual human activity.

Measuring the World

With the recognition of the irrational numbers, suddenly shapes such as triangles, squares and circles could be described. These were very important concepts when trying to measure distances or estimate movements of the planets. A new way of using numbers was therefore developed: numbers to represent lines and shapes made from lines.

Lines and shapes had always been important to ancient civilizations, particularly for the marking out of territories and fields. This was a much more serious problem for the Egyptians, for the Nile had a habit of flooding every year and wiping out all of the farm and plot boundaries. The careful division of the land into fresh boundaries each year became known as *geometry*, which comes from the Greek words *geo,* meaning Earth, and *metrein*, meaning measure. When the ideas of using numbers to define lines and shapes spread, so did the word, and today geometry is only associated with lines and polygons.

Lines and simple shapes were so commonly used (try drawing anything without them) that many of the ideas were explored by mathematicians who lived as long ago as 800 B.C. The Pythagoreans studied shapes such as triangles as they investigated and proved the Pythagorean theorem. One important

Above: Jacques-Louis David's painting The Death of Socrates *depicts Socrates in his final moments of life, as he reaches for the cup of hemlock.*

mathematician who was probably influenced by the Pythagoreans was a man named Hippocrates born at around 470 B.C. in Chios, Greece (not the same Hippocrates who became a doctor — this was a mathematician). Hippocrates of Chios is not remembered very favorably. Aristotle tells us that Hippocrates first worked as a merchant but was too stupid to look after his money and was swindled by custom-house officers at Byzantium. Hippocrates also may have believed that light originated in the eyes of the observer, and tried to explain comets and the Milky Way as optical illusions caused by moisture oozing out of nearby planets and stars. But whether lacking in common sense or not, Hippocrates was the first to write a serious book about geometry, which he called *Stoicheia*, meaning *Elements of Geometry*.

He was not the last. In 427 B.C., a man called Aristocles was born in Athens, Greece. He became known for his nickname of "Plato," meaning broad. It is not known whether the name refers to his broad shoulders (apparently created by his expertise in wrestling), his broad forehead or his broad range of abilities. Plato spent time in military service and politics, and probably befriended Socrates, for his uncle was a close friend of the famous Greek philosopher. Plato gave up his ambitions to make a difference as a politician when his 70-year-old mentor Socrates was executed in 399 B.C. on charges of "corrupting the youth" and impiety (meaning a lack of respect for God and not doing your duty). Although the 27-year-old Plato was very upset by this, it seems that Socrates was well-known as an outspoken opponent of democracy and may have slept with some of his male students. Certainly it seems that he had a relationship with the son of one of his accusers which, combined with his unrepentant attitude at the trial, probably helped lead to his death.

Nevertheless, Plato was dismayed by the death of his friend and left Athens to travel around Egypt.

Eventually, in 387 B.C. Plato returned to Athens and set up a school of higher education on land owned by a man called Academos. He named it the *Academy* after the landowner and the word academy has been used to mean an exclusive place of higher education ever since. The Academy was influenced by Pythagoreans in Italy, and taught mathematics as a branch of philosophy and religion. One example of this is the set of Platonic solids, where it was thought that the elements (earth, fire, air and water) were comprised of geometric atoms. Earth was thought to be composed of cube-shaped atoms, fire of tetrahedrons, air of octahedrons and water of icosahedrons. The fifth Platonic solid, the dodecahedron, was thought to be the shape of the universe.

Below: Examples of the five Platonic solids as depicted in Davisson's engravings in Philosophia Pyrotechnia *(Paris, 1642).*

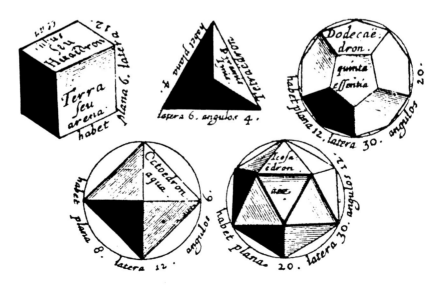

Plato made many important contributions to philosophy and his belief that mathematics was the best training for the mind was to influence scholars for hundreds of years. Above the doors of the Academy were the words, "Let no one unversed in geometry enter here."

Plato was very influential, but unfortunately his writing was not terribly clear. He often wrote his ideas in the form of conversations, and seemed to believe that this was one of the most important ways to learn. If you were to study at the Academy, the course was an arduous 15-years long. Students spent the first 10 years studying sciences and mathematics such as plane and solid geometry, astronomy and harmonics. They then had to spend a further five years studying the more difficult subject of "dialectic" — the art of conversation, of question and answer. By the end, students were trained to question and find answers about the fundamental nature of things. Their aim was to ground all knowledge on unbreakable truths.

Although there might be few people willing to study in the Academy today, Plato had several students who became well-known mathematicians, making some important contributions to geometry. Plato's Academy lasted for a remarkable 900 years, finally being closed down in 529 A.D. by the Christian Emperor Justinian who believed it was a pagan establishment.

But perhaps the most successful and important work in geometry was performed by Euclid (our friend from Chapter 1 who invented the Fundamental Theorem of Arithmetic — the one about prime numbers being "inside" all other numbers). It turns out that Euclid's 13-volume book may not have been all his own work. Today it is widely believed that Hippocrates of Chios is responsible for some of the contents of the first couple of books, and Plato's students are responsible for some of the other content.

The reason why Euclid's work was so groundbreaking was not that he had invented everything himself, it was that he wrote down all the current mathematical findings in such a clear way. In his 13-volume book he defined clearly, for the first time, many of the fundamental concepts needed in geometry and some of the axioms on which much of our mathematics is based. Many of his definitions may seem quite silly today, but they are still valid and very important. For example, in one axiom Euclid tells us that "things that are equal to the same thing are equal to each other." This may seem perfectly obvious: if I have the same number of apples as my friend Jon, and you have the same number of apples as Jon, then you and I have the same number of apples as each other. Or to write it using a little math notation:

If $a = c$ and $b = c$ then $a = b$.

But of course it is important that the axioms are clearly true, otherwise all our mathematics would no longer work properly. There's not much point in creating a fundamental truth that is sometimes not true. For example, what about: "no sum of two numbers can be greater than their product." This sounds plausible: $2 + 3$ is clearly not greater that 2×3. But unfortunately, this "truth" quickly fails: $1 + 3$ is greater than

*Above: Raphael's School of
Athens (1509). Plato and
Aristotle stand in the middle
of the painting.*

1 x 3, and it fails even more miserably with
fractions and negative numbers. So part of the
genius of Euclid was in figuring out which truths
were really true.

Euclid also defined what points and lines are.
For example, he told us that "one can draw a
straight line from any point to any point." In
other words given any starting point, and any
ending point, you can draw a straight line from
the first to the second. He went on to tell us
that all right angles were equal, and that two

Left: Euclid's definition of lines and points innovated geometry for generations to come. In this painting, Portrait of Nicholas Kratzer, *by Hans Holbein the Younger, German mathematician Kratzer is depicted experimenting with angles.*

parallel lines will never cross each other. This led to what became known as Euclidian geometry — the regular and logical math of shapes that has been used ever since. In Euclidian geometry, a square is always made from four right angles, and a shape drawn in one location will be the same shape if it is moved somewhere else. But that doesn't mean Euclid was right. Actually, he wasn't — and non-Euclidian geometry needed to be invented to explain why. It wasn't until Einstein figured out some bizarre things about space and time before we really understood how wrong Euclid was — and also why gravity screwed up the math (which we will explore in a later chapter). Nevertheless, Euclid was close. His assumptions and proofs were good enough for us to continue to use Euclidian geometry without any problems for all the designs, machines and buildings we've ever made since.

Moving the World with Numbers

Geometry quickly became an essential tool for mathematicians, scientists and engineers. Perhaps the best and earliest examples are the achievements of Archimedes. Today he is known for his legendary cry of "Eureka!" and his naked streak through the streets as he blindly celebrated his discovery about the displacement of water in his bath. But Archimedes was actually obsessed with geometry.

Archimedes was born a decade before Euclid died. He was primarily a mathematician and was friendly with the successors of Euclid in Alexandria. He even used to send them copies of his latest mathematical theories — which he soon discovered they were passing off as their own. But rather than become bitter and angry (as later mathematicians such as Bernoulli did), it seems that Archimedes found the whole thing amusing. He decided to try and catch them out by sending them two fake (and deliberately flawed) theories to see if they would still claim ownership. He relates this story in the preface of his book *Spirals*, saying that he did this:

> ... so that those who claim to discover everything, but produce no proofs of the same, may be confuted as having pretended to discover the impossible.

Clearly Archimedes was a clever fellow, and he was actually quite a famous mathematician in his lifetime. Unlike others such as Pythagoras who saw numbers as being steeped in mystical symbolism,

Archimedes was inspired by mechanisms and geometry that he observed around him. He explained in his book called the *Method:*

> ... certain things first became clear to me by a mechanical method, although they had to be proved by geometry afterwards because their investigation by the said method did not furnish an actual proof.

Archimedes was obsessive about geometry, to the extent that he even forgot to bathe. According to a Roman writer:

> Oftimes Archimedes' servants got him against his will to the baths, to wash and anoint him, and yet being there, he would

Below: Portrait of Archimedes.

ever be drawing out of the geometrical figures, even in the very embers of the chimney. And while they were anointing of him with oils and sweet savours, with his fingers he drew lines upon his naked body, so far was he taken from himself, and brought into ecstasy or trance, with the delight he had in the study of geometry.

The practical nature of Archimedes was apparent through his many inventions. He invented the spiral pump — today known as the Archimedes Screw — which draws water up a pipe by rotating a cleverly shaped spiral within it. This device is still used to pump water around the world today and uses a

similar principle to that used by propellers on boats and aircraft to push air and water. Archimedes also played a substantial role in defending his home during the siege of 212 B.C. as the Romans attacked Sicily. He had been persuaded to design some extraordinary ship-crushing machines by his friend and relation King Hieron II of Syracuse, Sicily. We know a surprising amount of detail about these ancient war machines, for a Roman called Plutarch wrote a biography of one of the Roman commanders who fought against Archimedes. According to Plutarch:

These machines [Archimedes] had designed and contrived, not as matters of any importance, but as mere amusements in geometry.

But what may have been mere amusements to Archimedes became known as legendary

Below: Illustration of the Archimedes Screw.

machines. Plutarch's account of the battle sounds like something out of an action movie:

Above: Archimedes invented formidable machines that were used to defend Sicily against the invading Romans, including machines that supposedly lifted ships into the air by an iron hand or crane's beak.

> ... when Archimedes began to ply his engines, he at once shot against the land-forces all sorts of missile weapons, and immense masses of stone that came down with incredible noise and violence; against which no man could stand; for they knocked down those upon whom they fell in heaps, breaking all their ranks and files. In the meantime huge poles thrust out from the walls over the ships and sunk some by great weights which they let down from on high upon them; others they lifted up into the air by an iron hand or beak like a crane's beak and, when they had drawn them up by the prow, and set them on end upon the poop, they plunged them to the bottom of the sea; or else the ships, drawn by engines within, and whirled about, were dashed against steep rocks that stood jutting out under the walls, with great destruction of the soldiers that were aboard them. A ship was frequently lifted up to a great height in the air (a dreadful thing to behold), and was rolled to and fro, and kept swinging, until the mariners were all thrown out, when at length it was dashed against the rocks, or let fall.

These extraordinary weapons become more believable when we discover that Archimedes also invented the lever and the compound pulley. It is said that he demonstrated the pulley for his friend Hieron by showing how a fully loaded ship could be moved by one man. Archimedes is also reported as saying "Give me a place to stand and rest my lever on, and I can move the Earth." So perhaps tipping a few Roman ships over in the middle of a battle was not such an impressive feat for him.

When Is a Number not a Number?

Geometry was clearly an important tool for those wanting to draw and design with accuracy. But it took a thousand years before it became the flexible tool we know today. It was a man named Abu Ja'far Muhammad ibn Musa al-Khwarizmi, born in Baghdad around 780, who invented a significant new way of using numbers in mathematics. Al-Khwarizmi was a scholar in the House of Wisdom — an academy where ancient Greek philosophical and mathematical texts were translated (such as the work of Archimedes and Euclid). Al-Khwarizmi probably learned from these texts, and became well known as a mathematician after writing his book: *Hisab al-jabr w'al-muqabala.* (Take a look at the words in that title and see if you can spot anything familiar.) He was a practical mathematician who explained that his work would teach useful skills to normal people:

> ... what is easiest and most useful in arithmetic, such as men constantly require

Above: Russian stamp with portrait of al-Khwarizmi, a well-known mathematician and scholar who worked on equations relating to shapes.

in cases of inheritance, legacies, partition, lawsuits and trade, and in all their dealings with one another, or where the measuring of lands, the digging of canals, geometrical computations and other objects of various sorts and kinds are concerned.

Al-Khwarizmi concentrated on solving equations related to shapes. He didn't use any of the notation we use today; he used words to explain the problem and pictures to solve it.

Solving equations related to shapes

"... a square and 10 roots are equal to 39 units. The question therefore in this type of equation is: what is the square which when combined with 10 of its roots will give a sum total of 39?

The manner of solving this type of equation is to take one-half of the roots just mentioned. Now the roots in the problem before us are 10. Therefore take 5, which multiplied by itself gives 25, an amount which you add to 39 giving 64. Having taken then the square root of this which is 8, subtract from it half the roots, 5 leaving 3. The number 3 therefore represents one root of this square, which itself, of course, is 9. Nine therefore gives the square."

Today we would write the same problem as:

$$x^2 + 10x = 39$$

What is the value of x?

Al-Khwarizmi's explanation was perhaps not the clearest ever written, but thankfully it's not that difficult to see what he was talking about. His geometric proof is elegant, for he showed that you only need to draw some squares in order to work out the value of x.

First we need to draw the right shapes. We begin with a square in the middle with sides of length x. This has area $x \times x$, or x^2. To draw rectangles that have a total area of $10x$, we need to add four rectangles of sides x and 10/4, because 10/4 x 4 x x = $10x$. The problem tells us that these five shapes have a total area of 39.

Now we add four smaller squares of sides 10/4 to make one big square. We know the area we added: it's 4 x 10/4 x 10/4 = 25. So we know the area of the big square must be 39 + 25 = 64. To find the length of one side of the big square we need to know the root of the square (the square root) of 64, which is 8 because 8 x 8 = 64. Finally, from the diagram we know that the length of the side is 10/4 + x + 10/4 = 8. To put it another way:

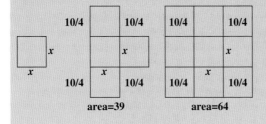

area=39   area=64

$x + 5 = 8$. So the value of x must be 3.

Al-Khwarizmi relied on this kind of geometry to figure out the unknown elements in an equation. He used two methods to simplify his equations and make them easier to draw: *al-jabr*, meaning "completion" and *al-muqabala*, meaning "balancing." These are what we recognize today as algebraic operations, for example *al-jabr* allows us to transform $x^2 = 40x - 4x^2$ into $5x^2 = 40x$. Alternatively,

al-muqabala allows us to reduce $50 + 3x + x^2 = 29 + 10x$ into $21 + x^2 = 7x$.

Over the centuries, we lost the second term and now refer to this kind of mathematics as *algebra*. We use letters of the alphabet to represent unknown numbers, and we apply the normal arithmetic operators to them as though they were numbers. This is a wonderful extension to mathematics. Suddenly we can describe and manipulate numbers even when we don't really know what they are. Algebra gives us the mathematical equivalents of the vague word "thingy." So we can now say, "the amount is twice thingy" and then try and figure out what number "thingy" is (except that mathematicians usually write "the amount is $2x$" and try to figure out the value of x).

Al-Khwarizmi's geometric proofs were one of the first methods of figuring out which numbers the letters might actually be; there are many other methods. Algebra is now one of the most widely used tools in mathematics. You have probably noticed that algebra has already been used several times in this and previous chapters — it's such an important invention that it's hard *not* to use. It also forms a central part of computer programming, which uses *variables* to do most computation — they work just like the letters of the alphabet in algebra except that whole words are often used instead (so we really can use words like "thingy" as variable names). Al-Khwarizmi's invention of algebra makes him one of the most influential mathematicians of all time.

An Equation Paints a Thousand Words

Algebra was a tremendously important invention because it allowed us to write down numbers that couldn't be written down. At last the problem of irrational numbers became irrelevant. All you had to do was say $x = \sqrt{2}$ and suddenly you could manipulate x like any other number.

But it took another 800 years after Al-Khwarizmi before a French philosopher and mathematician realized that algebra could also be used to define geometric shapes as well as numbers.

René Descartes was born in La Haye, France, in 1596 (now named Descartes, after him). Descartes studied philosophy and mathematics for many years. (During this time he suffered from ill health and was given permission to sleep until 11 a.m. every morning — a habit he continued for almost all of his life.) After finishing his studies, Descartes traveled extensively around Europe, finally settling in Holland. He began work in physics and mathematics although was slightly nervous of publishing his work after the imprisonment of Galileo by the Inquisition. Despite his worries, he produced a treatise on his scientific research, which had three appendices on optics, meteorology and geometry. The work on optics was nothing terribly new, and much of the work on meteorology was wrong (for example he believed that water that had been previously boiled would freeze more quickly). But the work on geometry was groundbreaking. One of his most important contributions was the combination of algebra with geometry to produce analytic geometry. He had the clever idea that if a letter could represent a number, then two letters x and y might be able to represent a point in space, and several letters might be able to represent a line, or a circle or any other shape. From Descartes we were given Cartesian coordinates: the two letters (x, y) tell us that a point is a distance x along a horizontal direction (known as the x axis) and a distance y up a vertical direction (the y axis).

Using algebra to define geometric shapes

Descartes also explained how we could write a straight line as $y = mx + c$. In English, this means: if you know the gradient (steepness) of the line is m, and you know that the line crosses the vertical y axis at c, then you can plot the vertical distance y for every horizontal distance x of the line.

So for the equation $y = 3x - 1$, if we put in three values of x: 1, 2, 3

$y = 3 \times 1 - 1 = 2$

$y = 3 \times 2 - 1 = 5$

$y = 3 \times 3 - 1 = 8$

The equation of the line gives us three points on the line (1, 2), (2, 5) and (3, 8). If you have the Cartesian coordinates of anything, then you can draw it by joining the dots:

But the clever thing about analytic geometry is that you don't need to draw it to work things out. For example, if we wanted to figure out where this line crosses the x axis, we could draw it and take a guess from the picture, or we could simply say: what is the value of x when $y = 0$? The rules of algebra then allow us to find the answer:

$0 = 3x - 1$

$3x = 1$

$x = 1/3$

So the point we're interested in is: (1/3, 0)
In the same way, we can use equations to define curved lines, such as

$y = x^2$

or even circles, like this:

$(x - h)^2 + (y - k)^2 = r^2$

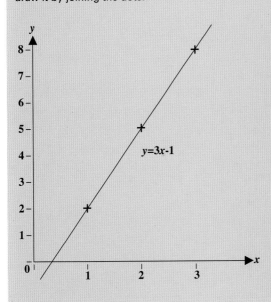

Today this form of geometry has become an essential tool for science and engineering. It is much quicker and more accurate to use algebra to calculate results about geometric forms than to draw them.

Descartes was both a philosopher and a mathematician. He loved mathematics, as he considered it to be the only subject where truth could be known absolutely. As well as his famous saying, "I think, therefore I am," he also said, "With me everything turns into mathematics." Descartes also had a sense of humor. Today we may remember him for his contributions to philosophy and analytic geometry (every time we use the word "Cartesian" we remember his name). But Descartes had his own wishes on how we should remember him:

I hope that posterity will judge me kindly, not only as to the things which I have explained, but also to those which I have intentionally omitted so as to leave to others the pleasure of discovery.

Above: A page from Portraits des Grands Hommes, *illustrated by Desfontaines depicting Descartes at his desk.*

Lost in the Margins

Descartes was not always so cheery. One man he not only argued with, but tried to discredit was a French lawyer called Pierre de Fermat. Fermat studied mathematics whenever he could and made several contributions to geometry, although he rarely wanted to publish his results. He began corresponding with other famous mathematicians of the day, but made the mistake of reviewing the work of Descartes as "groping about in the shadows." From that moment, Descartes loathed Fermat and even when Fermat's math was shown to be correct when his was wrong, Descartes still did his best to ruin the reputation of Fermat.

As we saw in Chapter 1, Fermat was interested in number theory, rediscovering a pair of amicable numbers. But today he is remembered most for what he did not write

Fermat's last theorem

Fermat was referring to an equation that derived from the famous Pythagorean theorem. If you recall, the Pythagorean theorem is a simple way to calculate a side of a right-angle triangle, whereby if a, b and c are the lengths of the sides, then $a^2 + b^2 = c^2$. Fermat had already proved that if the values of a and b were rational numbers then c had to be irrational. His note in the margin referred to a very similar equation:

$$a^n + b^n = c^n$$

Fermat's last theorem states that for values of n greater than 2 it is impossible to find integer solutions to the equation. This is a surprising idea, for we can clearly find solutions to the equation when $n = 2$. For example, $a = 3$, $b = 4$ and $c = 5$ is one solution because

$$3^2 + 4^2 = 5^2$$

and of course it becomes the familiar Pythagorean theorem once again.

But make n equal to 3 or more and there is no way to find these solutions. We just can't do it.

down. The discovery was made after his death when his son Samuel published a translation of Diophantus' book *Arithmetica* (another important early work on algebra) with his father's notes. One of the notes that Pierre Fermat had scribbled in the margin was "I have discovered a truly remarkable proof which this margin is too small to contain."

So Fermat's mysterious little note suggested that not only did he think this theorem was true, but he had proved it was always true. He just didn't have the space in the margin to write it down. For the next 300 years, mathematicians were aggravated by this casual little comment, because no one else could figure out how to prove it. Had this French lawyer somehow derived a remarkable proof? He certainly had a habit of not recording all of his work properly. But why could hundreds of other mathematicians not discover it?

Today it is widely believed that if Fermat did think of a proof, it was almost certainly wrong or incomplete. We think this because in 1994, a British mathematician called Andrew Wiles finally managed to prove Fermat's last theorem, after spending much of his career worrying about the problem. His proof is 150 pages long.

Somewhat bizarrely, the story of Fermat's last theorem had so captured the imagination of mathematicians that after its proof had been found it had the unique distinction of being made into a musical, called *Fermat's Last Tango*. Descartes would not have been pleased, although perhaps he would have appreciated the irony of Fermat accidentally achieving exactly what he wished to: a kind judgement in posterity because of an omission.

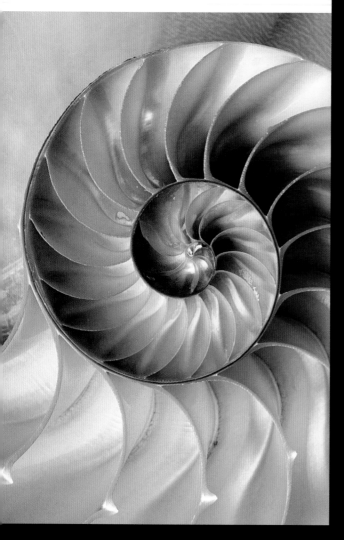

The discovery of irrational numbers did not stop people from believing that numbers lay at the heart of existence. In fact, the nature of irrational numbers — mysterious unknowable numbers that seemed to describe many geometrical shapes — gave rise to the idea that there might be special numbers inside the universe even more attractive. Perhaps there was an irrational number that described shapes all around us. Maybe God was leaving clues by using this magical number again and again, in the shapes of life.

GOLDEN PHI

CHAPTER φ

Many philosophers and mathematicians believed in a magical number for hundreds of years. It still is the belief of some people today. Why? Because they think they know one of these magical irrational numbers. Today we call it phi, or φ (pronounced "fi" to rhyme with eye). Its value is approximately 1.6180339887498948482 … (but like all irrational numbers, it goes on forever, never forming a repeating pattern).

Phi can be found in many measurements and ratios of lengths of ancient Greek sculptures, architecture and even the Egyptian pyramids. Some claim that the human body is made from ratios equal to phi, and that it lies at the heart of all that is beautiful and pleasing to the eye. Today this number is considered so special that it is called the golden section, golden ratio or simply, the golden number.

Above: Leonardo da Vinci's painting of the Mona Lisa *is thought to employ the principles of the golden ratio.*

Seriously, Rabbits?

Phi is so crucial in the definition of some forms (just as pi is so crucial in the definition of circles as we will see in a later chapter) that it is likely that its appearance in ancient structures and artworks is more coincidence than intentional. Some modern-day philosophers claim that Plato was aware of phi and even incorporated it into his philosophical thinking. But unfortunately, Plato often wrote in cryptic riddles and so he may not have been aware that the mysterious value of phi

Opposite: The nautilus shell presents one of the finest natural examples of a logarithmic spiral.

lurked behind some of his mathematics. Others claim that Leonardo da Vinci used the golden ratio to help him find the perfect proportions for his famous painting of the *Mona Lisa*. There may be more truth to this tale, for it is known that da Vinci was taught mathematics and illustrated a book in 1509 called the *Divina Proportione* (*The Divine Proportion*) by mathematician Luca Pacioli, which was all about the golden ratio. Pacioli certainly believed that this number was special, writing in his book that phi:

> … just like God cannot be properly defined, nor can be understood through words, likewise this proportion of ours cannot ever be designated through

Above: Fra Luca Pacioli with Mathematical Instruments *by Jacopo de Barbari.*

intelligible numbers, nor can it be expressed through any rational quantity, but always remains occult and secret, and is called irrational by the mathematicians.

The very first investigations of phi may have been made some 2,000 years earlier by the heretical Pythagorean Hippasus (the one who suspiciously drowned) or by a colleague of his called Theodorus. But it was our friend Euclid who first wrote down the definition of how to find phi. (If you recall, Euclid was the guy who wrote the 13-volume book containing the Fundamental Theory of Arithmetic and Euclidian geometry.)

Nearly 1,500 years after Euclid was born, another mathematician called Leonardo was born in Pisa, Italy. The son of Guglielmo, or Bonaccio as he was called (meaning good-natured), Leonardo became known after his death as Fibonacci (short for *filius* Bonacci, or "son of Bonaccio").

Because of his father's work, Fibonacci was educated in North Africa and came to understand the new Arabic numerals and positional numbering system in use there. He quickly realized that the use of the symbols 0 to 9 were far superior to the Roman numerals

Euclid's ratio

Euclid doesn't use the term "golden ratio," which is much more recent. But he does explain how to calculate its value. Given a line from point A to point B, the golden ratio is formed when a point C is chosen on the line so that the ratio of distances AB:AC equals AC:CB.

Euclid also describes how this ratio can be found inside many geometric shapes. For example, given a pentagon, if a line is drawn from every corner to every other corner, the lines cross each other at the golden ratio. Once again, the ratio of distances AB:AC equals AC:CB (but it works for all the other crossing lines as well).

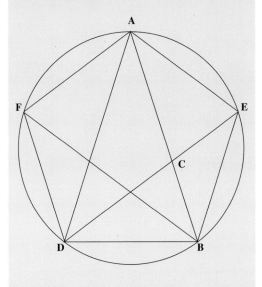

Above: Leonardo Pisano, also known as Fibonacci, founder of modern mathematics, is well known for Fibonacci numbers and the Fibonacci sequence.

still in use in Europe, and so became influential in bringing this new form of number notation to Europe. He did this quite intentionally, by writing a book called *Liber Abaci*, or *Book of Calculation* (although in those days it meant *Book of the Abacus*). He wrote it for tradesmen rather than academics, giving examples of how to write the numbers, how to calculate profits, losses, convert between currencies and calculate interest. He also included a number of mathematical problems, and (probably to his great surprise, had he known) it is for one of these problems that Fibonacci is remembered best. This is the puzzle that made him famous:

> A certain man put a pair of rabbits in a place surrounded on all sides by a wall. How many pairs of rabbits can be produced from that pair in a year if it is supposed that every month each pair begets a new pair which from the second month on becomes productive?

Despite also writing puzzles about spiders crawling up walls and hounds chasing hares, it was this puzzle about rabbits that captured the imagination of people for centuries to come. The reason why can be found in its solution:

The Fibonacci sequence acts like a light that steadily illuminates more and more of phi. The larger the numbers in the sequence, the more we see of the true value of phi. We know that phi is irrational, so (by definition) there do not exist two integers that when divided

Fibonacci sequence

The puzzle says that we start with one pair of rabbits and each pair of rabbits becomes mature in the second month and able to "beget" a new pair. How many pairs of rabbits are there? If we write out the answer for each month, we get:

1, 1, 2, 3, 5, 8, 13, 21, 34, 55, 89, 144, 233, ...

To see why, imagine you're standing among the rabbits. We begin with 1 pair. After one month, we still have 1 original pair. After two months we have the original plus a new pair, making 2 pairs. After three months we have the original pair, the new pair and another new pair produced by the original, making 3 pairs. After four months we have the previous 3 pairs, plus we have another 2 pairs produced by those older than one month, making 5 pairs. And so it goes on ...

It should be immediately clear that this sequence has a pattern to it. Each number is made by adding the two previous numbers together.

The sequence became known as the Fibonacci sequence. It's special because of what happens when you divide each number with the one before it in the sequence:

$\frac{3}{2} = 1.5$

$\frac{5}{3} = 1.666...$

$\frac{8}{5} = 1.6$

$\frac{13}{8} = 1.625$

$\frac{21}{13} = 1.61538...$

$\frac{34}{21} = 1.619047...$

$\frac{55}{34} = 1.617647...$

$\frac{89}{55} = 1.61818...$

Now flick back a page or two and take a look at the value of phi again. Do you notice anything? Try dividing the next few numbers in the sequence with each other and see what happens.

Equiangular spiral

The equiangular spiral is known as a
logarithmic spiral, which was named by
Jacques Bernoulli, brother of the dodgy
Johann who we met in Chapter 0. (We'll see
more of logarithms in a later chapter.) The
spiral can be drawn by dividing a rectangle
that has a width over length equal to the
golden ratio into squares and drawing a
quarter circle with radius equal to the side
of each square within that square.

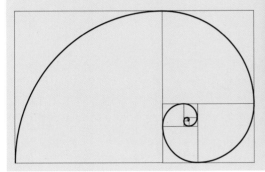

*Above: Illustration of Jacques
and Johann Bernouilli
discussing geometry.*

will give the precise value of phi. However,
the Fibonacci sequence produces numbers that,
when divided, forever get closer and closer
to phi.

Fibonacci wrote several other important
books on mathematics, but his work was largely
forgotten for centuries. He made contributions
to geometry and number theory, but today is still
remembered mostly because of those rabbits.

Phi is not only found in patterns of
reproducing rabbits and pentagrams. Descartes,
the philosopher and creator of Cartesian and
analytic geometry, was the first to notice a
special kind of spiral called the equiangular

spiral. Just as we saw with the pentagram
earlier, this spiral also contains phi, and so has
led many mathematicians, biologists and
philosophers over subsequent centuries to make
comparisons with spirals found in nature, for
example on seashells or snail shells. The closer
you look at natural forms, the more of these
spirals you see. There have been entire books
filled with illustrations of plant shapes, patterns
of petals and seeds, as well as spirals of shells
that point out where phi can be measured,
again and again. Yet more evidence, it is
claimed, that phi is a number that is fundamental
to life.

IOANNIS KEPPLERI
Mathematici Cæfarei
hanc Imaginem
ARGENTORATENSI BIBLIOTHECÆ
Confecr.
MATTHIAS BERNEGGERVS
k.al. Ianuar. Anno Chr.
M DC. XXVII

*Above: Astronomer Johannes
Kepler. This oil painting (1627)
hangs in the Musee de
l'Oeuvre Notre Dame,
Strasbourg, France.*

Out of this World

In 1571, 400 years after the birth of Fibonacci,
another mathematician was born in Württemberg,
of the Holy Roman Empire (now known as
Germany). Johannes Kepler was a deeply religious
man. His beliefs were then considered a little

known as the golden ratio or golden mean and perhaps the triangles of the Pythagorean theorem are much more suggestive of a cut gemstone. He wasn't to know, however.

Kepler studied Greek, Hebrew and mathematics at university, and in his first year received top grades for all except mathematics. That did not stop him from beginning what was to become a very successful career in astronomy and mathematics. Kepler's most important work attempted to explain the movement of the planets. He was one of the first to adopt the radical new Copernican system, in which instead of a belief that there were six planets (counting the Moon) that all orbit the Earth, it was believed that there were six planets (including the Earth) orbiting the Sun, with only the Moon orbiting the Earth. Kepler's approach to calculating the paths of the planets was surprisingly Platonic. He thought five geometric solids, each placed inside the other could perfectly explain their orbits.

unusual (eventually resulting in him becoming excommunicated), for he thought that phenomena such as the movement of planets could be explained by important numbers and geometric forms that had mystical significance. Perhaps not surprisingly, Kepler believed phi was especially important, referring to it as "the division of a line into extreme and mean ratio" and saying:

> Geometry has two great treasures: one is the theorem of Pythagoras; the other, the division of a line into extreme and mean ratio. The first we may compare to a measure of gold; the second we may name a precious jewel.

Kepler's metaphor would have worked a bit better the other way around since phi became

Right: An early representation of The Copernican World System that assumes circular orbits of the planets around the Sun and provided Kepler with the basis for his studies of planetary orbits.

Kepler's solution to the mystery of the cosmos

Kepler thought that a planet must follow a circle around the Sun. Or more precisely, planets acted as though they rolled around big invisible spheres that had the Sun in the middle. By drawing a sphere of the right radius around the Sun (where the radius was the distance from the planet to the Sun), he could figure out where the planet would move to next. His problem was to figure out how far from the Sun each planet was (or how big each sphere should be). He did this by using Plato's five regular geometric shapes to define the spacing between each planet.

He first drew a sphere for the path of Saturn, the outermost planet. Inside Saturn's sphere he drew a cube so that the corners just touched the sphere. Inside that cube he drew another sphere so that its edges just touched the inner sides of the cube. Jupiter followed the path of this sphere. Inside Jupiter's sphere he drew a tetrahedron, and inside that he drew another sphere. Mars rolled around the path of that sphere. Inside Mars' sphere he drew a dodecahedron, and inside this 12-sided solid he drew another sphere. Earth followed this path. Inside Earth's sphere he drew an icosahedron, and inside this 20-sided shape he drew another sphere. Venus followed this spherical path. Finally, inside Venus' sphere he drew an octahedron, and inside that he drew the final sphere, for Mercury.

Kepler was very pleased with this work as it seemed to tie everything together: Plato's solids, real observations of the movements of the planets, and even phi and the right-angled triangles of the Pythagorean theorem, which occur repeatedly within the Platonic solids. Euclid had proven that there can only be five regular convex solids, and these five shapes seemed to act as perfect spacers between the paths of the six planets. Kepler believed that this was clear evidence for the existence of a God who planned his creation around mathematics.

He wrote his work in his first book called *Mysterium cosmographicum* (meaning *Mystery of the Cosmos*). For a while it seemed as though he had solved that mystery. His model using the Platonic solids matched the real movements of the planets with extraordinary accuracy. The largest error was less than 10 percent, which is pretty good for a model even today.

Above: The model created by Kepler that demonstrated his theory of planetary orbits.

But to Kepler this was not good enough. He wanted to have a perfect model for then he might understand more about how and why the planets moved as they did. He continued his research, studying the orbit of Mars in enormous detail and also investigating optics and telescopes. He soon realized that despite the seeming elegance of his original ideas, they could not be correct, for his observations showed that the orbit of Mars followed an ellipse and not a circle. This was one of the earliest recorded examples of what we now call

Kepler's laws

Kepler realized that all the planets followed elliptical orbits around the sun (a fact that we now call "Kepler's first law"). He also realized that the speed of the orbit depended on how close the planet was to the Sun. Planets actually speed up as they zip past the Sun, and their changes in relative speed can be calculated by chopping the elliptical path in segments of equal areas. So when the planet is farthest away from the Sun, draw a line from the planet to the Sun, and an hour later draw another line to make one segment. Now, wherever the planet is in the future all we need to do is draw another line from it to the Sun and the second line must be drawn such that the resulting segment has the same area — and the second line will point to where the planet must be an hour later. When the planet is close to the Sun this means that it has to move much quicker in order to make a fat segment with the right area. When it's further away, it moves slower to make a thin segment with the same area. We now call this idea "Kepler's second law."

Sometime later, Kepler found a third law, which states that for any two planets, the ratio of the squares of their periods will be the same as the ratio of the cubes of the mean radii of their orbits. Another way of writing this is:

$$\frac{P1^2}{P2^2} = \frac{R1^3}{R2^3}$$

Where $P1$ is the time it takes for planet 1 to orbit the Sun (the length of its year), $P2$ is the time it takes for planet 2 to orbit the Sun, $R1$ is the average distance of planet 1 from the Sun, and $R2$ is the average distance of planet 2 from the Sun.

This clever little equation tells us that as we move farther from the Sun, the length of the year of the planets increases rapidly (the planets orbit much slower). Using this equation we can figure out exact times and distances for the planets. For example, if we say $P2$ is the Earth, which takes one year to orbit and is an average distance of 1 astronomical unit (or 92,900,000 miles/149,508,057 km) from the Sun, and if we know that Mercury is an average distance of 0.3873 au, then:

$$\frac{P1^2}{1^2} = \frac{0.3873^2}{1^3}$$

So working it out, $P1$ must be the square root of 0.0580955, which is 0.241 of an Earth year. So Mercury takes 88 days to orbit the Sun.

Perihelion   **Ampelion**

Equal areas in equal times

"observational error," where the accuracy of an idea is checked against observations of reality in order to confirm or refute that idea. This process is now central to science, and is how we are able to check the validity of our explanations of phenomena around us. Kepler was a good enough scientist to abandon his idea when the data clearly showed he was slightly wrong, despite having written a book and built his early career on that very idea. Eventually Kepler figured out mathematical rules that did describe the motion of the planets properly.

Kepler's laws were an impressive achievement, especially since he had no understanding that the cause of the motion of the planets was gravity. Newton figured that out a few years later and was able to improve Kepler's laws and make them even more accurate (we'll see more of Newton in a later chapter).

Below: Johannes Kepler is shown discussing his discoveries of planetary motion with his sponsor, Emperor Rudolph II, a firm supporter of science and the arts.

Above: Oil painting of the Moon and its inhabitants as described in Kepler's Somnium.

One final, and lesser known of Kepler's achievements was as the author of perhaps the first ever science fiction story. Toward the end of his life he wrote a book called *Somnium* (*The Dream*). In this fictional tale he described how a student was transported to the Moon with the assistance of a Daemon. His imagination was remarkable, suggesting that the departure from the Earth would be traumatic for the student, "for he is hurled just as though he had been shot aloft by gunpowder to sail over mountains and seas." This sounds as though Kepler was thinking that a large rocket might be needed (recall this is before gravity was understood at all and long before the invention of aircraft). He also says that once the speed has become large enough, "we are carried along almost entirely by our will alone, so that finally the bodily mass proceeds toward its destination of its own accord." This sounds like the concept of inertia. Kepler had somehow figured out that the trip to the Moon through space requires acceleration up to the right speed with a rocket, then that speed will be maintained until deceleration takes place — exactly how the actual lunar landers reached the Moon in the 1970s. He also described a large number of other difficulties faced by his hero during the journey, showing that he had seriously thought such a journey was possible, albeit arduous.

Once the student had arrived on the Moon, Kepler used his story to explain the motion of the planets, correctly anticipating that an observer on the Moon would see the Earth rise and set in a very similar manner to the way we see the Moon from the Earth. He also made some guesses about life that might live on the Moon. He thought creatures inhabiting the Moon would grow to monstrous size and would be nomadic as no towns were visible by telescope on the Moon.

> Some use their legs, which far surpass those of our camels; some resort to wings; and some follow the receding water in boats; or if a delay of several more days is necessary, then they crawl into caves. Most of them are divers; all of them draw their breath very slowly; hence under water they stay down on the bottom.

This must be the first ever description of how life might be on another world. Again, Kepler's imagination is remarkable as he invents new types of creatures that are suited to his notions of lunar landscape, especially since he wrote this hundreds of years before Darwin and the theory of evolution.

It is very fitting, then, that NASA has named a new spacecraft after Kepler. The Kepler Mission comprises a powerful space telescope to look for Earth-sized planets around other stars. The spacecraft is scheduled to launch in 2008. Johannes Kepler would have been delighted.

Don't Be Absurd

Another way of calculating the value of phi is to add 1 to the square root of 5 and divide by 2. It's very hard to write this down accurately because the square root of 5 is another irrational number that goes on forever. If we write it down to 10 decimal places, then our calculation of phi will only be accurate to 10 decimal places. Because of this (as we saw in the last chapter) we typically write:

$(1 + \sqrt{5}) / 2$

In mathematics, when we get a bit stuck, we use a different way of writing down something. We can't write down irrational numbers made from taking the square root of some numbers. One option is to use a letter such as x to represent these numbers and then use algebra to manipulate them. But algebra is really for numbers that we don't know the values of. The irrational result of a square root has a value that we (more or less) do know. We just can't write it down. The solution is to use the square root symbol and write the number as a *surd*.

The word "surd" began life meaning the same as "irrational." It seems that Arabic translators in the ninth century translated the Greek word *alogos* (irrational) to *asamm* (deaf, dumb). The Arabic mathematicians liked to think of rational numbers as being audible, and irrational as inaudible. Their word was later translated into Latin and became *surdus* (deaf, mute).

Today, surds are considered to be irrational numbers that we can't write down properly except like this:

$\sqrt{5}$

As soon as mathematicians think of a new way of writing numbers, a crowd of others immediately

1.61803_3

Rules of surds

Rule 1: $\sqrt{ab} = \sqrt{a}\,\sqrt{b}$

For example, $\sqrt{12} = \sqrt{4}\,\sqrt{3} = 2\sqrt{3}$

Rule 2: $\sqrt{(a/b)} = \sqrt{a}\,/\,\sqrt{b}$

For example, $\sqrt{(3/4)} = \sqrt{3}\,/\,\sqrt{4} = \sqrt{3}\,/\,2$

Rule 3: $a\sqrt{b} + c\sqrt{b} = (a+c)\sqrt{b}$

For example, $5\sqrt{5} + 4\sqrt{5} = 9\sqrt{5}$

In fact, surds are just one form of notation. We also commonly use indices, which mean:

We write "x squared" like this: $x^2 = x \times x$

We write "the reciprocal of x squared" like this: $x^{-2} = 1\,/\,(x \times x)$

We write "square root of x" like this: $x^{1/2} = \sqrt{x}$

And we write "cube root of x" like this: $x^{1/3} = \sqrt[3]{x}$

There are even more rules about how indices interact. I bet you thought writing down a number was easy …

think of new mathematical rules that apply. Not surprisingly then, we can also manipulate surds using a set of rules.

Surds are not terribly popular today, despite the efforts of Simon Stevin, a Belgian born in 1548, 23 years before Kepler. Stevin helped introduce decimal fractions into Europe and made a strong plea in his work that all types of numbers, whether fractional, negative, real or surd should be treated equally as numbers.

Stevin was a very practical man, becoming quartermaster-general of the army of the States-General (north Netherlands) and advising on the building of windmills, locks and ports. He also devised a method of flooding lowlands in the path of an invading army by opening sluicegates in dikes. Stevin was very successful in mathematics, writing 11 books on subjects as diverse as hydrostatics, music and astronomy. He even discovered three years before Galileo that objects of different weights fall at the same speed. He discovered this by dropping two lead balls, one 10 times heavier than the other, at a height of 30 feet from the church tower in Delft.

The surd is no longer a popular representation of numbers, regardless of the plea of Stevin and despite the fact that a surd lies within the golden number phi. In practical applications we now use indices (or the inaccurate decimal approximations) to represent these irrational numbers, mainly because they are easier to represent with computers.

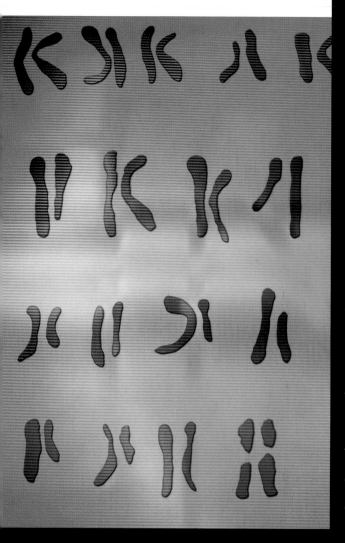

There are a lot of numbers between one/two (which is why Chapter 2 is the seventh chapter of this book). But a couple of millennia ago, two was considered to be the first number. One is simply a unit, but two is the first number to mean "a collection of things." Two is also the first even number we've seen so far. It's the first even number there is, for the definition of "evenness" is "exactly divisible by two."

GOOD AND EVEN

CHAPTER 2

To be even has always meant more than a dull mathematical definition. An even number can be divided exactly into two halves, so it is in balance. Two is therefore the most crucial of even numbers. This belief has been captured by many religions and philosophies. The Chinese believe that the number two captures the essence of the two polar forces of *yin* (the receptive, constrictive female energy) and *yan* (the creative, expansive male energy). Even though two is the first of the *yin* numbers, when spoken in Cantonese it sounds the same as "yi," meaning "easy." So the number two is often put in front of other lucky numbers to improve them, for example, 23 sounds like "easily growing," 26 sounds like "easily profitable" and 29 sounds like "easily enough" when spoken in

Above: The diabolic number two. A devil from this detail of The Damned in Hell by Luca Signorelli from the fresco cycle in the San Brizio Chapel (Capella della Madonna di San Brizio) in the Orvieto Cathedral.

Cantonese. But 24 is very unlucky, for it sounds the same as "easily dying." The number 2,424 is very unlucky indeed.

Early Christians thought two represented the Devil, or the division between soul and God. The Zoroastrians believe that two is symbolic of an eternal, evenly balanced battle between good and evil. And in Russia, if you're going to send flowers to someone, make sure you send an odd number. Even numbers are for funerals. There are also many other superstitions, some sensible, some quite bizarre. For example, it's supposed to be unlucky to have two holes in the same sock, but lucky if two people sneeze simultaneously. It's supposed to be unlucky if two people pour tea from the same pot,

Above: Bust of Baron
Gottfried Willhelm Leibnitz,
the mathematician who
developed the philosophy
of binary numbers.

There Are 10 Types of People In the World: Those Who Understand Binary and Those Who Don't

Computers are the world's best number manipulators. But ironically, the number 2 is never used inside a computer. Computers don't use any other numbers greater than 2 either. The only numbers that our computers ever manipulate are 0s and 1s. The reason why is because their numbering system is in base 2, or binary, instead of base 10, or decimal, as we are used to. Binary is really just a different way of writing down numbers. All the same numbers are still there, it's just that they're written down (or stored in computer memory) using lots of 1s and 0s.

The origins of binary are ancient, but perhaps the first person to study binary in detail was the son of Catharina Schmuck. Gottfried Leibnitz was born in 1646 in Saxony (now Germany). He had a profound interest in philosophy and poetry, which he wrote in Latin. He also had wide and varied interests from politics to inventing to mathematics. From 1678–79 he designed a number of wind-operated pumps and gears to drain water from the mines in the Harz mountains, but could never actually get any built. He'd also been working on an automatic calculating machine that could perform multiplication by doing repeated additions, but it took him over 20 years to make a model that worked. Leibnitz was an extraordinary letter-writer, corresponding with over 600 mathematicians, philosophers, engineers and politicians as he developed his varied ideas. He developed a friendship with Johann Bernoulli (the argumentative fellow we met in Chapter 0) but managed to have long-lasting feuds with the likes of Newton over who

but lucky if two shoots grow from the root of a single cabbage. There are several more superstitions about eggs as well: find an egg with two yolks and there may be a death in your family; but break two eggs accidentally and you will find your soulmate. Let's just hope your soulmate doesn't mind you dropping eggs everywhere.

Counting in binary

It takes four binary digits or "bits"
to count to 15:

0000	0
0001	1
0010	2
0011	3
0100	4
0101	5
0110	6
0111	7
1000	8
1001	9
1010	10
1011	11
1100	12
1101	13
1110	14
1111	15

Whatever the base, numbers are still used in the same way: there are just fewer (or more) digits. So in base 10, we use 10 symbols 0 to 9 and the first four digits of any number represent 10^3, 10^2, 10^1 and 10^0, or thousands, hundreds, tens and ones. In base 2, we use only the two symbols 0 and 1 and the first four digits represent 2^3, 2^2, 2^1 and 2^0, or eights, fours, twos and ones.

(Look back to Chapter 0.000000001 to see how binary counting resembles the beads on an abacus.)

was the first to think of several mathematical ideas (including the invention of calculus, which we'll see more of in a later chapter). It may be that his tact did not match his inventiveness, for in response to a mathematician named Keill who had accused him of plagarism, he said that he "refused to reply to an idiot."

Despite his tumultuous relationships with people, Leibnitz made many important discoveries in mathematics. His development of binary became something of a philosophy. He believed that the universe could be represented more elegantly through binary and its conflicting, yes-no, on-off nature such as male-female, light-dark, right-wrong. He proposed that life and thought could be reduced into a series of binary propositions and began obsessively translating numbers into seemingly never-ending lists of binary 1s and 0s. Toward the end of his life he began to believe that binary numbers represented Creation, with 1 symbolizing God, and 0 depicting Void.

Weaving Patterns of Numbers

The on-off nature of binary was noticed a few years later by a French silk weaver and inventor. Joseph Jacquard was born in Lyon, France, in 1752. When his father died he inherited two looms and rather unsuccessfully tried to continue the business of weaving. He was unsuccessful because he was more interested in improving the design of the looms than using what he had to

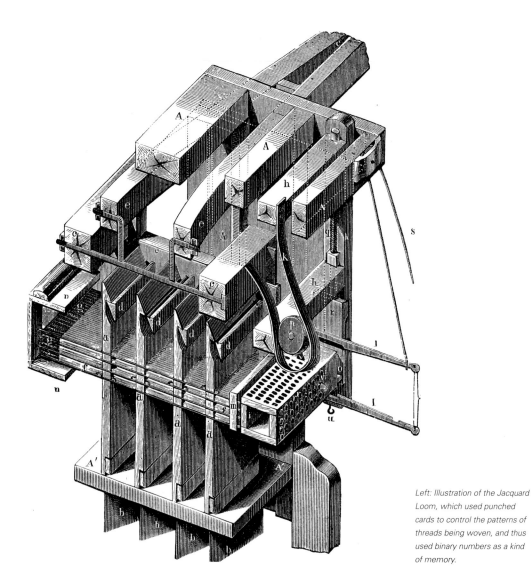

Left: Illustration of the Jacquard Loom, which used punched cards to control the patterns of threads being woven, and thus used binary numbers as a kind of memory.

make money. Eventually he had to give it up to become a limeburner, before fighting in various wars (and losing his son, who was shot down at his side). When he returned from war, Joseph worked in a factory and spent his spare time designing and constructing his improved loom.

His design was revolutionary, for it allowed a loom to be "programmed" using large pasteboard cards with holes in them that controlled the patterns of threads to be woven. Suddenly even intricately patterned fabrics could be woven by anyone — much to the dismay of the silk-weavers who vehemently opposed the invention. But the advantages of the Jacquard Loom were too great to be suppressed and the invention was declared public property, with Jacquard receiving a large sum of money and a royalty on all future Jacquard Looms. This made him a wealthy man — by 1812 there were 11,000 looms in use in France.

Although Jacquard's loom was purely mechanical and did not perform any mathematical

Above: Joseph Marie Jacquard,   *Right: Charles Babbage, creator*
inventor of the Jacquard Loom.   *of the Difference Engine.*

operations, its punched cards are the first example of using binary numbers as a kind of memory. By punching holes in the cards, Jacquard was permanently storing binary digits, using a hole to represent a 1 and no hole to represent a 0. This was an idea that was to prove useful.

Thirty-nine years after Jacquard was born, a lad called Charles Babbage was born in London, England. Charles suffered from several childhood illnesses and was packed off to Devonshire to be looked after and educated by a clergyman. He was later sent to private schools by his wealthy father and educated further at home, so that by the time he began at Trinity College, Cambridge, he found the mathematics course very easy. He was taught all the important mathematical ideas of the time, including those of Newton and Euler, and was particularly impressed with Leibnitz's ideas. Even as an undergraduate, Babbage achieved some impressive feats, for example

setting up a society to translate important foreign works and even publishing a history of calculus (which included a description of the feud between Newton and Leibnitz).

Babbage married at the age of 22 and moved back to London. Before long he was publishing his own works in mathematics and was elected as a fellow of the Royal Society and several other prestigious societies, some of which he helped create. This was despite his rather low opinion of the Royal Society, for according to his writing:

> The Council of the Royal Society is a collection of men who elect each other to office and then dine together at the expense of this society to praise each other over wine and give each other medals.

(Some would say not much has changed in the following 200 years.)

Babbage was not the most impressive mathematician ever known. He published a few ideas, but unfortunately some were wrong. But

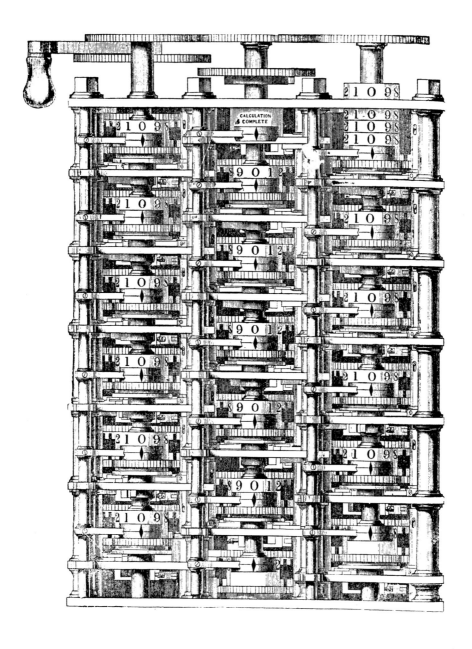

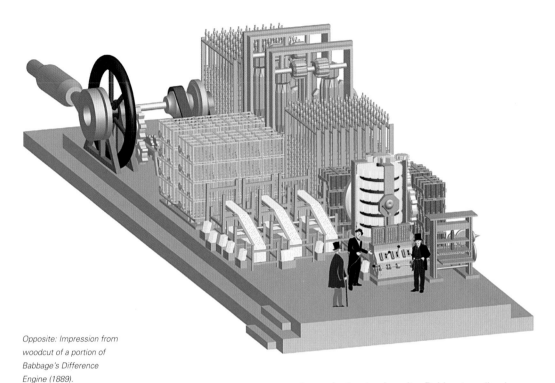

Opposite: Impression from woodcut of a portion of Babbage's Difference Engine (1889).

Right: Artist's impression of the scale of Babbage's steam-powered Difference Engine

Babbage became famous for an idea he had when only 20 years old — a machine to perform mathematics automatically. He would later call this machine the Difference Engine, for it produced tables of numbers by calculating and adding differences to the previous numbers. This was potentially enormously useful, for many complex calculations (for example, in astronomy, design or even artillery firing) relied on tables of numbers to speed up the math. It was much quicker to figure out a product of two huge numbers by looking it up on a table, than to calculate it afresh each time. But when all such tables were calculated by hand, errors were common — meaning that anyone using the table would find their calculation wrong. While still an undergraduate at university, Babbage realized that a machine might be able to produce far more accurate tables, quicker. By 1822, at the age of 30, Babbage had completed a working prototype of a difference engine (see box, page 94).

Babbage's Difference Engine impressed so many people that he was given several large sums of money by the British government to design and build a much larger version. This was to have six orders of differences (compared to just two as we saw before) and would calculate numbers 20 or more digits in length. But the project became a monster, taking years and sucking thousands of pounds of government (and Babbage's own) money. The Difference Engine Number 2 was never completed. Until, that is, 200 years after the birth of Babbage, when the

London Science Museum constructed a fully working Difference Engine according to his plans, complete with automatic printer. You can go and see it today. And it does work.

Although Babbage struggled to have his Difference Engine constructed, he continued his research. Eventually he realized that a mechanical computing machine could theoretically do more than simply produce tables of numbers. Babbage realized that a machine could be made to calculate anything. He had conceived of a general purpose computing machine that could be programmed to compute anything that the user required. His designs showed that this would have been a goliath of a machine: 100 feet (30 m) long, 30 feet (10 m) wide and powered by a steam engine. It would be programmed using the same punched cards used in the Jacquard Loom, and would be capable of printing its own punched cards, as well as a printer to output numbers, a curve plotter to draw curves and a bell to announce when computations were complete. Although entirely mechanical, this was truly a computer. It would have executed programs in the way modern computers do, and its programs allowed loops, conditional branches and arithmetic operations, meaning that it could have performed any calculation possible. Babbage called it the Analytic Engine and said:

Babbage's Difference Engine

Using nothing more than cleverly shaped cogs and wheels, Babbage's first Difference Engine was capable of calculating a series of numbers using two orders of differences. In other words, given a starting number, it would use the second difference to make the first difference, and it would use the first difference to make the next number in the sequence.

So if the first difference was 0, the second difference was 2 and the starting number was 41, then the machine would iteratively calculate the new numbers:

0 2   41

2 2   43

4 2   47

6 2   53

8 2   61

As you can see, the second difference (the second number) is added to the previous first difference (the first number) each time, and the resulting first difference is then added to the previous third number each time.

Using only successive additions in this way, Babbage had demonstrated that a machine could calculate successive terms of the equation: $n^2 + n + 41$

It was not terribly quick, managing about 60 new numbers every five minutes, but it was more reliable than a human could ever be.

As soon as an Analytical Engine exists, it will necessarily guide the future course of the science.

But it never did exist beyond his designs and scientific papers. The breakthrough was beyond the current technology. Like the simpler Difference Engine, no one could build it. Unlike the Difference Engine, no one has dared to build it in modern times. Nevertheless, Charles Babbage is still regarded as the father of the computer, for when the first electronic computers were constructed over a hundred years later, they worked according to a design that was remarkably similar to his. They even used punched cards to enable the input and output of binary numbers.

Thinking Logically

Binary has always been at the heart of computers, and not just because binary numbers are easy to imprint on punched cards. The on-off, true-false nature of binary is also central to logic, and computers are very logical devices.

The son of a shoemaker, George Boole was born in 1815. It was clear from an early age that Boole had a very special brain. His father taught him mathematics and asked a friend to teach him Latin, but could only afford to send him to a local commercial school. So George taught himself Greek — to such a high standard that the local schoolmaster refused to believe that a 14-year-old could translate from Greek so well and accused him of cheating. Two years later

Above: Portrait of George Boole.

Boole became a teacher himself, supporting his family after the collapse of his father's business. Astonishingly, by the time he was 19, George Boole had opened his own school in Lincoln. Four years later he took over running another school, and by the time he was 25 he started his own boarding school.

You would think running all these schools would be enough, but Boole was learning

Boolean logic

Boole realized that the logical operators AND, OR and NOT were all you needed to describe any logical statement you liked. This statement might be an English sentence, e.g., "I will take my umbrella when it's raining **AND** it's either overcast **OR** it's calm." Or the statement might be an electronic circuit, e.g., "My circuit Q will output 1 when input A is 1 **AND** either inputs B **OR** C are 1, otherwise it will output 0." Logically, the two statements are exactly the same.

As you might expect, because we have yet another way of writing down math, there are yet more rules about how we can manipulate what we've written down. These form Boolean algebra and they enable us to take any logical expression and simplify or transform it, while keeping its meaning exactly the same. One example of this can be seen when we draw a truth table and use it to create a corresponding Boolean expression. Returning to the example, if we're deciding whether to take our umbrella, let's call the decision Q. When Q is 1, we take it, when Q is 0 we leave the umbrella at home. Since we're worrying about three things, let's label them too: A (raining), B (overcast), C (calm). So for every possible value of A, B and C, there must be a value for Q (a decision about the umbrella). If we list them all, we need to count in binary:

A B C	Q
0 0 0	0
0 0 1	0
0 1 0	0
0 1 1	0
1 0 0	0
1 0 1	1
1 1 0	1
1 1 1	1

And now we can see that Q is 1 on only three occasions: when A = 1, B = 0, C = 1 or when A = 1, B = 1, C = 0 or when A = 1, B = 1, C = 1.

If we were a little silly, we could write that out and try to use it:

I will take my umbrella when it's raining **AND NOT** overcast **AND** calm **OR** when it's raining **AND** overcast **AND NOT** calm **OR** when it's raining **AND** overcast **AND** calm.

Which can be written in something a little closer to Boolean algebra as:

Q = (A and ~B and C) or (A and B and ~C) or (A and B and C)

But there are lots of handy simplification rules that allow us to simplify this expression step by step, until we reach:

Q = A and (B or C)

Which is much easier to understand. In fact, it's not only easier to understand, it's better to make into a circuit. In electronics, transistors are used to behave as AND, OR and NOT logical gates. The 1s and 0s are electrical currents (on or off). Computers are made from millions and millions of these transistor logic gates. But by using Boolean algebra we can simply use the logical expressions that define the circuits and reduce the number of gates we need. In the example, instead of using 10 logical operators, after simplification we only needed two, meaning fewer transistors and faster, more efficient computers.

mathematics at the same time. It was soon clear that — despite the lack of any university education — Boole was imaginative and highly original in mathematics. He became a Professor of Mathematics at Queen's College, Cork, when he was 34, where he stayed for the rest of his life, which sadly was only 15 years more. But it was long enough for him to make a breakthrough in mathematics that has borne his name ever since: Boolean logic.

Boole was considered by his colleagues as a genius. In the words of his colleague De Morgan:

> Boole's system of logic is but one of many proofs of genius and patience combined ... That the symbolic processes of algebra, invented as tools of numerical calculation, should be competent to express every act of thought, and to furnish the grammar and dictionary of an all-containing system of logic, would not have been believed until it was proved.

In 1864 he took his usual 2-mile walk from home to College and was soaked in a downpour. He lectured in his wet clothes and became ill, probably with pneumonia. In a strange twist of fate, his wife had her own slightly frightening logic. She believed that the cure of an illness was often the same as its cause, and so decided to throw buckets of cold water over him as he lay in his bed. Boole never recovered.

Knocking Down the Foundations of Mathematics

While binary numbers, punched cards, automatic calculating machines and Boolean logic all led to the design of modern computers, a discovery that threatened the whole of mathematics created the theory behind computers.

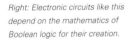

Right: Electronic circuits like this depend on the mathematics of Boolean logic for their creation.

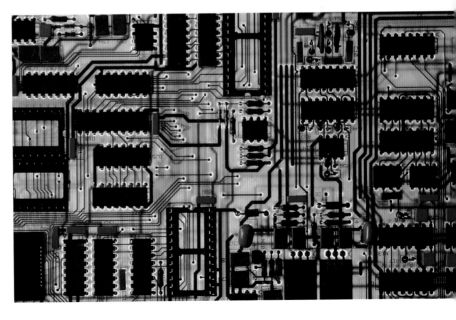

Left: Bertrand Russell receiving the Nobel Prize in Literature from King Gustav VI Adolph in Stockholm, Sweden (1950).

Eight years after the death of Boole, Bertrand Russell was born in Wales. Both his parents had died by the time he was 4, so he was brought up by his grandmother (against the legal will of his father who had wished him to be raised by two atheists). Russell was educated at Trinity College, Cambridge, in mathematics and the moral sciences. His morals and personal beliefs were to play a significant role throughout his adult life as he actively campaigned against the two World Wars, being dismissed from various jobs at universities and even spending time in prison because of his beliefs. But Russell also was awarded the Order of Merit, the Nobel Prize for Literature and, with his friend Albert Einstein, released the Russell-Einstein Manifesto in 1955, calling for the curtailment of nuclear weapons. Throughout his long life he maintained his strong beliefs and morals, whether popular with the current governments or not, which gave him some fame as a public figure.

Despite his strong morals, Russell's main work was in mathematics and logic. He was able to show that mathematics is reducible to logic — that is, all mathematical findings can be rewritten as logical expressions. This was great, for it helped us understand all those fundamental truths that mathematics is built upon. But then he discovered a paradox: something that was both true and not true at the same time. As we saw in Chapter 1, a proof by contradiction relies on this kind of thing — when something seems to be both true and false at the same time, it means that the reasoning must be faulty. But Russell's paradox seemed to imply that the whole of mathematics was faulty (see box, opposite). This was a complete disaster!

The reason why mathematicians found the whole thing so appalling was because it identified what seemed to be a flaw in the foundations of mathematics. Centuries upon centuries of mathematical ideas and proofs were all built upon a series of basic, fundamental truths. But Russell's paradox suggested that no proof could be trusted any more. The notion that mathematics was the only area where truth could be known absolutely, as Descartes had believed, was no longer valid.

This work caused a flurry of activity in an attempt to fix the problem. But rather than remove Russell's paradox, it got worse. In 1931 a mathematician proved once and for all that mathematics would always be incomplete. His name was Gödel.

Kurt Gödel was born in 1906 in Brünn, Austria-Hungary (now Brno, Czech Republic). He suffered from rheumatic fever as a child and

Russell's paradox

Russell's paradox is quite similar to the Barber's paradox. Think about this:

There is a barber who shaves precisely those people who don't shave themselves. Does he shave himself?

If he doesn't shave himself, then he must shave himself. But if he does shave himself, then he will not shave himself! The only way this makes sense is if he shaves himself and does not shave himself at the same time — but that's logically not possible. That's why it's a paradox.

Russell's paradox is similar, but it's about sets, or groups of things. Russell knew that if you can have, say, a set of cups, and a set of saucers, then you could have a set of sets of cups and saucers. In other words the concept of "set" is a useful mathematical notion, and we can group sets inside other sets. Many proofs of basic arithmetic operations such as addition and subtraction are made using ideas of sets of numbers, so they form some of the fundamental building blocks of mathematics. Russell knew that it is possible for some sets to be inside themselves. One example of this is

the set of all non-empty sets. If you have a set of anything, then it's in this set. Because there's something in this set, it is a non-empty set itself, and so must be included in the set of non-empty sets. So it's inside itself. Or, as we say in set theory, it's a member of itself.

So far so good. No paradoxes here, just some slightly weird ideas. But Russell thought of a very special kind of set, which was perfectly acceptable in mathematics yet made no sense at all. Russell's paradox asks:

There is a set of all sets that are not members of themselves. Is the set a member of itself?

This set will contain itself only if it does not contain itself. But if it does not contain itself, then it will contain itself. Like the Barber's paradox, the only solution that makes sense is if the set both contains itself and does not contain itself at the same time. But this is logically impossible. It's like being over 6 feet tall and under 5 feet tall at the same time — it can't happen.

when 8 years old read medical books about the condition. He was obsessive about his health from then on, believing he had a weak heart — an obsession not helped by his famous paralyzed, wheelchair-bound mathematics lecturer Furtwängler who used an assistant to write on the board as he taught.

Eventually Gödel took a position at the University of Vienna, but as World War II began, he found himself persecuted as a Jew (even though he was not Jewish) and also became fearful that he might be asked to fight. He fled to the United States with his wife and became a U.S. citizen in 1948. During his interview Gödel attempted to tell the judge that he had found a logical loophole in the United States Constitution that conceivably could allow a dictator to emerge — luckily his friends Einstein and Morgenstern calmed him down and the judge didn't listen.

Opposite: Einstein photographed handing the first Albert Einstein Award for Achievement in the Natural Sciences to Kurt Gödel.

Also pictured are Lewis Straus (center) and Julian Schwinger (right).

Gödel's most memorable work became known as Gödel's incompleteness theorems. The first and perhaps most celebrated theorem was this:

> For any consistent formal theory that proves basic arithmetical truths, it is possible to construct an arithmetical statement that is true, but not provable in the theory. That is, any consistent theory of a certain expressive strength is incomplete.

> Put simply, it means that we cannot prove everything using mathematics. Some truths are not possible to prove.

It was a devastating result, for it showed that the quest by hundreds of mathematicians over millennia could never succeed. It would never be possible to create a fully complete system of mathematics where everything from the lowest axioms to the highest, most complex proofs could be shown to be unequivocally true. It didn't matter how perfectly the foundations of mathematics were laid, there would always be some truths that could never be proved. Gödel's result was proved, so all the world could do was accept that mathematics was not infallible. Just as we cannot ever write down the complete value of an irrational number, sometimes it's not possible to prove something mathematically — and there's nothing we can do about it.

Gödel's obsession with his health never left him. His contributions in mathematics did not help him understand medicine, either. His brother, who happened to be a doctor, wrote:

> My brother had a very individual and fixed opinion about everything and could hardly be convinced otherwise. Unfortunately he believed all his life that he was always right not only in mathematics but also in medicine, so he was a very difficult patient for doctors. After severe bleeding from a duodenal ulcer ... for the rest of his life he kept to an extremely strict (over strict?) diet which caused him slowly to lose weight.

Toward the end of his life Gödel worried so much about germs that he wore a ski mask wherever he went and obsessively cleaned his eating utensils. He died at the age of 72, probably through malnutrition as he had been refusing to eat, believing he was being poisoned.

Above: Photograph of Alan
Turing, who helped crack
German Enigma codes during
World War II.

It Does Not Compute

Regardless of their personal beliefs, Russell and Gödel had begun a landslide in mathematics. Things were slipping away. What would turn out to be unprovable next? This was a question that attracted Alan Turing, a Londoner born in 1912. While investigating this question, he would invent the theory behind modern computers.

Turing was never very happy at school, often doing poorly while creating wildly original answers to mathematical problems. He infuriated his teachers by pursuing his own studies and performing his own experiments, and yet still managed to win almost every math prize at his school. Turing went on to study mathematics at King's College, Cambridge, learning of Russell and Gödel's work. When only 24 years old, after already producing some impressive work, he published his ideas on decidability and logic. Turing managed to prove that it was not possible to show universally (for any given examples) that a logical or arithmetic statement was true or not. This was yet another nail in the coffin of "perfect mathematics," but what was more important was the way in which Turing had constructed his proof. He had imagined a machine that would read a long tape, follow instructions it read, spool to different places on the tape and write

symbols back onto the tape. A strange tape reading, writing and spooling machine.

The Turing Machine was more than a mathematical flight of fancy. It was a conceptual machine capable of performing computable calculations. Turing also thought of the idea of a universal machine that could simulate the behavior of any other Turing Machine. He had proved such a universal machine existed and that it would be able to perform any possible computation, given the right instructions. This was what we needed — a true general purpose computing machine.

The Universal Turing Machine became the theoretical blueprint for all electronic computers in the world. It told us how computers needed to behave, and helped us design them and make them real. Because of this theory we have always known that any computer can perfectly simulate the behavior of any other computer (given enough time and memory). We've known this even before the first electronic computers were made, because of Turing.

Turing never saw the computerized world he helped shape. He continued working at Cambridge and at Princeton, before being recruited in 1938 to work for the government on a top-secret code-breaking project. When World War II broke out, Turing began working full-time at Bletchley Park, where he developed imaginative and brilliant methods for breaking the German Enigma codes used to send messages to the Luftwaffe and the German navy. It has been said that his efforts may have saved more military lives during the war than

those of any other.

After the war, Turing returned to Cambridge, then took a position at Manchester University where he continued his research into computers. He had created a design for an electronic computer for the National Physics Laboratory in London, but quickly moved on to consider amazingly advanced topics such as artificial intelligence and patterns of growth caused by interacting chemicals. Turing regarded biology and human brains as computational devices and according to one writer:

… he became involved in discussions on the contrasts and similarities between machines and brains. Turing's view, expressed with great force and wit, was that it was for those

The Turing Machine

Turing's peculiar imaginary tape machine became known as a Turing Machine. He used it to show that some problems were undecidable in mathematics. He did this by imagining that this little computing machine was performing a calculation, following the symbols on its tape. He then asked the question: is it possible to tell if this machine will get stuck in an endless loop and calculate forever, or if it will stop calculating and give an answer? It would be quite possible for it to calculate forever, for example, if its tape said at point A, "spool to point B" and at point B it said, "spool to point A."

Turing figured that if it was possible to tell if his machine halted or not, then another machine should be able to do it, for he knew that his imaginary machine could theoretically do any mathematical calculation. So he imagined a second Turing Machine that would

examine the first and halt, outputting "won't halt" if the first would never halt, or just running forever if the first machine did halt.

Now for the clever bit. Turing imagined what would happen if the second machine looked at itself, and tried to decide if it would stop calculating or not. Suddenly there was a paradox: if the machine ran forever then it would stop; but if it stopped then it would run forever. This is logically impossible and so proves that there exist some Turing Machines that are undecidable — we will never be able to tell if they halt or not. Although this may seem like a very obscure and unlikely situation, it turns out there are a very large number of undecidable or uncomputable problems — a fact that has been causing problems for computer programmers ever since.

Above: Bletchley Park, Buckinghamshire, 1926. Headquarters of Allied cryptographs, where German Enigma and Lorenz codes were deciphered.

who saw an unbridgeable gap between the two to say just where the difference lay.

In 1943, at the Bell Labs cafeteria he was once heard to say in his high-pitched voice:

No, I'm not interested in developing a powerful brain. All I'm after is just a mediocre brain, something like the President of the American Telephone and Telegraph Company.

Even today, the Turing test is the most well-known example of an intelligence test for computers. It works like this: imagine you're

When Turing thought about these ideas there was no Internet, computers were the size of a room and not even as powerful as a modern pocket calculator. His foresight was remarkable.

In 1952, Turing was arrested for homosexual behavior, which was illegal in Great Britain at that time. Following this, his security clearance at Bletchley Park was removed and the government became suspicious that he was a security risk. Sadly, Turing died in 1954, aged only 42, of cyanide poisoning while conducting electrolysis experiments. The cyanide was

sitting in front of a computer, chatting with two people online. If you can have a wide-ranging conversation with both people without being aware that you are really talking to one person and one computer, and without being able to distinguish between the two, then the computer can be called intelligent and would pass the Turing test. To date, some computer programs have managed to pass — but only when the subject is confined to a very narrow topic. No program has passed the Turing test when there is no limit to the topic being discussed.

Below: Enigma coding machine used by Germans during World War II.

found on a half-eaten apple beside him. It is widely believed he committed suicide, but his mother insisted it had been an accident.

There's a popular myth that the computer manufacturer, Apple, had their half-eaten apple logo inspired by this event. In reality, the original logo depicted Newton with an apple over his head, and was nothing to do with Turing.

Above: Photograph of Robert Oppenheimer and von Neumann with early computer (1952).

Designing Computer Architectures

Turing had laid the theoretical foundations of computers and Babbage had designed the first mechanical computer, but it took a genius to design one of the first electronic computers ever built. Born in 1903 in Budapest, Hungary, János von Neumann quickly became a child with extraordinary abilities. His father would show him off at parties, for he could memorize a complete page of a phone book in seconds, then correctly recite names, addresses and phone numbers. He soon showed amazing abilities in mathematics at school, but at the request of his father, who wanted him to study something he could make a living from, he chose to study chemistry at the University of Budapest. In his spare time he would attend mathematics courses, where his genius was quickly recognized. One mathematician who taught him, later said:

> Johnny was the only student I was ever afraid of. If in the course of a lecture I stated an unsolved problem, the chances were he'd come to me as soon as the lecture was over, with the complete solution in a few scribbles on a slip of paper.

Von Neumann continued his studies, doing a doctorate in mathematics and continuing with post-doctorate research at various universities.
By his mid-20s he was famous among established mathematicians as a "young genius" and soon moved to Princeton, where he became a professor, joining Albert Einstein and several other prominent mathematicians in the newly founded Institute for Advanced Study. Unusually, his genius did not prevent him from enjoying a full social life — the parties of Johnny von Neumann were legendary. His achievements for the rest of his life are far too

numerous to mention, but among his many mathematical works was an investigation of enormously complex equations to describe hydrodynamics (predicting the flow of water). He soon realized that the equations needed some form of computer to help in their analysis, and so drew up designs for the EDVAC — one of the first electronic computers that was ever built. His designs were similar to those of Babbage, but were designed to be implemented using electrical valves instead of mechanical cogs. The EDVAC had four logical elements: the Central Arithmetical unit (CA) where all the work was done, the Central Control unit (CU) which determined what would happen next, the Memory (M) for storing numbers and Input/Output devices (IO) such as keyboards and

Right: Technical Director T. Kite Shaepless demonstrating the EDVAC computer.

printers. The von Neumann architecture, as it became known, has been used as a template for almost every computer since.

Von Neumann didn't just design the conventional computer, he also created cellular automata — a kind of parallel computer that is used to analyze and model many complex systems even today. He happened to supervise a certain Alan Turing during his doctorate, when Turing came to visit in 1936–38. And like Turing, von Neumann also made contributions to the war, working on the mathematics and physics that led to the development of the hydrogen bomb.

Despite his genius — or more likely because of it — von Neumann did not react well to the news he was dying of cancer at the age of 52. In the poignant words of those who knew Johnny:

> When von Neumann realized he was incurably ill, his logic forced him to realise that he would cease to exist, and hence cease to have thoughts ... It was heartbreaking to watch the frustration of his mind, when all hope was gone, in its struggle with the fate which appeared to him unavoidable but unacceptable ... his mind, the amulet on which he had always been able to rely, was becoming less dependable. Then came complete psychological breakdown; panic, screams of uncontrollable terror every night.

His friend Edward Teller said,

> I think that von Neumann suffered more when his mind would no longer function, than I have ever seen any human being suffer.

Von Neumann's sense of invulnerability, or simply the desire to live, was struggling with unalterable facts. He seemed to have a great fear of death until the last ... No achievements and no amount of influence could save him now, as they always had in the past. Johnny von Neumann, who knew how to live so fully, did not know how to die.

Von Neumann lost his struggle in 1957 at the age of 53. His work lives on in all the computers of the world, but his first love was always mathematics. In his own words:

> If people do not believe that mathematics is simple, it is only because they do not realize how complicated life is.

Creating the Information Revolution

As we live our lives in our modern computer-filled world, we tend to forget the machinery that lies underneath it all. We ignore the several billion electronic devices (that von Neumann helped design) that permeate our environments. We can afford to ignore them, because we know they all do an equivalent job — some may be faster, some may use clever tricks or more memory, but they

Above: Photograph of Dr. Claude E. Shannon with his electronic mouse at the Bell Telephone Laboratories. The mouse has a *"super" memory and can learn its way around the maze after just one try.*

are all Universal Turing Machines. Rather than think about computers, we instead spend almost all of our time worrying about information.

Information may seem like an enormously obvious concept in a world where a tiny portable phone can take a digital photograph and email it to a friend who can put it online where anyone in the world can download it. The information revolution has genuinely transformed our world, with economies increasingly becoming dependent on electronic commerce over the

Internet, governments relying on the Internet to communicate with its people, and a huge range of communication gadgets allowing us to call or see who we wish, wherever we are on the planet. In 2004 the computer game Everquest was reported to have an economy so large that it became the 77th richest country of the world — even though Everquest is a virtual software world and does not exist in any physical form at all. Today, some of the richest and most successful companies either sell software or operate entirely online (such as Microsoft, eBay and Google).

It took another genius to tell us what information was. Claude Shannon was born in Gaylord, Michigan, in 1916. He was not as outstanding in mathematics as the likes of von Neumann, but instead focused on practical ideas such as electrical relay-based computers in his studies at Massachusetts Institute of Technology (MIT), before taking a job at the research laboratory of AT&T Bell Telephones in New Jersey. There he made his breakthrough on the subject of information, coining the name "bit" for binary digit, and showing that a bit forms the fundamental unit of information. He showed that the simplest possible 2-state on-off binary numbers could represent anything. This was an impressive insight in a time when electronic computers were still being designed and built. Through his work we learned how to use other

bits to perform error-correction and make sure information is not lost during transmission. We also learned how to compress information — which has led to compressed formats such as MP3 for music, digital television, and JPEG for digital pictures.

Shannon eventually returned to MIT as a professor, where his inventiveness and playfulness was clear for everyone to see, for he had a habit of unicycling down the corridors while juggling. At one point he worked on a motorized pogo-stick, and he also invented a two-seater unicycle and other variations. According to his colleagues:

> No one was ever sure whether these activities were part of some new breakthrough or whether he just found them amusing … A later invention, the unicycle with an off-centre hub, would bring people out into the corridors to watch him as he rode it, bobbing up and down like a duck.

Shannon continued his research (in between inventing bizarre forms of transport), also examining artificial intelligence and creating chess-playing programs. He created one of the first autonomous computer-controlled robots — a robot mouse that could navigate through a maze. Toward the end of his life, as he saw how computers and the information they generated were changing the world, he modestly said:

Information theory has perhaps ballooned to an importance beyond its actual accomplishments.

Shannon died in 2001, suffering from Alzheimer's disease in his final years.

Today we understand that information is the blood that flows within computers. The electronic computing machinery helps generate, transform, store and push this information around the huge networks of the Internet and wireless telephone networks. Information — which is nothing more than unthinkably vast numbers of binary 1s and 0s — flows day and night. That flow is becoming larger and larger as computers become faster and networks are given larger capacities. Not so long ago we all used modems with computers, and could only send or receive about 48,000 bits a second. Now we have 1Mbit or even 16Mbit broadband networks, allowing up to 16 million bits a second to flow. Information is one of the driving forces in our modern world. Information used to mean only numbers, then we realized it also meant text. Now it means audio, video and soon three-dimensional forms. In the future, who knows?

There used to be a saying that time equals money. No longer. Today information equals money. Numbers in base 2 flowing at the speed of light rule the world.

Run-length-encoding

One simple example of a compression technique, used in fax machines, is called run-length-encoding. Imagine you're faxing a black and white image. If we look at the page one tiny dot at a time, scanning left to right, top to bottom, we can write down a long list of values. A "b" would mean there is a black dot there, a "w" would mean that it's white. Most documents have much more white page than black (whether an image or text). So a list of bits for every black or white dot would have an awful lot of "w"s, perhaps something like this:

wwwwwwwwwwwwbwwbwwwwwbwwbwwwwwww
wwwwwbwwwwwbwwwwbwwbwwwwwwwbbwww
wwwwwwww

Run length encoding replaces all those repeated elements with a little code. So to replace 8 "w"s in a row, we can write "8w" instead, which uses only two characters instead of eight. Using this idea, the example above could become:

10wbwwb4wbwwb11wb4wb3wbwwb5wbb10w
(Of course the information is stored as binary numbers not letters like this, but the idea is the same.) The fewer digits we need to represent the same information, then the faster we can transmit it and the more we can store — which means faster faxes and emails, or more songs on your MP3 player and more digital channels on your TV.

69	83885	129	11059	189	27646	249	39620
70	84510	130	11394	190	27875	250	39794
71	85126	131	11727	191	28103	251	39967
72	85733	132	12057	192	28330	252	40140
73	86332	133	12385	193	28556	253	40312
74	86923	134	12710	194	28780	254	40483
75	87506	135	13033	195	29003	255	40654
76	88081	136	13354	196	29226	256	40824
77	88649	137	13672	197	29447	257	40993
78	89209	138	13988	198	29667	258	41162
79	89763	139	14301	199	29885	259	41330
80	90309	140	14613	200	30103	260	41497
81	90849	141	14922	201	30320	261	41664
82	91381	142	15229	202	30535	262	41830
83	91908	143	15534	203	30750	263	41996
84	92428	144	15836	204	30963	264	42160
85	92942	145	16137	205	31175	265	42325
86	93450	146	16435	206	31387	266	42488
87	93952	147	16732	207	31597	267	42651
88	94448	148	17026	208	31806	268	42813
89	94939	149	17319	209	32015	269	42975
90	95424	150	17609	210	32222	270	43136
91	95904	151	17898	211	32428	271	43297
92	96379	152	18184	212	32634	272	43457
93	96848	153	18469	213	32838	273	43616
94	97313	154	18752	214	33041	274	43775
95	97772	155	19033	215	33244	275	43933
96	98227	156	19312	216	33445	276	44091
97	98677	157	19590	217	33646	277	44248
98	99123	158	19866	218	33846	278	44404

In 2004, the company behind the massively successful Internet search engine Google announced their intention to raise funds for their future expansion. Rather than give the usual round figure of $1 billion or US$1.5 billion, they told the world that they intended to raise US$2,718,281,828. Why such a specific number? It was a mathematical joke – for this number is known in mathematics as e.

69	83885
70	84510
71	85126
72	85733
73	86332
74	86923
75	87506
76	88081
77	88649
78	89209
79	89763
80	90309
81	90849
82	91381
83	91908
84	92428
85	92942
86	93450
87	93952
88	94448
89	94939
90	95424
91	95904
92	96379
93	96848
94	97313
95	97772
96	98227

THE GREATEST

CHAPTER e

Google was also behind another use of e. The company put a mysterious message on billboards around the United States. It read:

{first 10-digit prime found in consecutive digits of e}.com

Those who could solve the puzzle and visited the resulting Internet site, discovered an even harder puzzle. Eventually, if all the puzzles were solved, the result was an Internet page containing a job advert for the brightest and best people to join Google.

e is one of those mysterious irrational numbers that lie at the heart of mathematics. Its value (to the first 20 decimal places) is 2.71828182845904523536. It's never been as popular as the golden ratio phi, or the number behind circles, pi. It didn't cause murders like the irrational √2 and it didn't lead to the invention of computers like binary 2. But e is hugely important and its discovery and use led to some of our most significant mathematical ideas. Before the invention of computers, we relied on e to help us perform accurate calculations. Without e, it would have been impossible to make the same progress in science and

technology over the last few centuries. If we had never found e, we might still be struggling to make complex machines that worked; we might have no cars or passenger airplanes. We might have had no computers. Instead of reading this book, you might have been working in a mill or coalmine.

INVENTION

Calculating Without Calculators

The story of e begins with a man called Neper, who was never to know that his work had anything to do with a new mathematical constant.

Jhone Neper — or to use the modern spelling, John Napier — was born in 1550 in Edinburgh, Scotland, to a wealthy family. Napier was educated in St. Andrews University, Scotland, and also Europe where he studied theology. By the time he was 24, Napier and his wife moved to a freshly built castle and he devoted his time to running his large family estates. Inventing and mathematics became his hobbies. When not studying theology he experimented with new ways to improve agriculture, using salts to improve soil. He also invented a mathematical tool, which he called the logarithm. It was a way of making complicated calculations much easier. In his words (written in 1614 and translated to English from Latin two years later):

Seeing there is nothing (right well-beloved Students of the Mathematics) that is so troublesome to mathematical practice, nor that doth more molest and hinder calculators, than the multiplications, divisions, square and cubical extractions of great numbers, which besides the tedious expense of time are for the most part subject to many slippery errors, I began therefore to consider in my mind by what certain and ready art I might remove those hindrances. And having thought upon many things to this purpose, I found at length some excellent brief rules to be treated of (perhaps) hereafter. But amongst all, none more profitable than this which together with the hard and tedious multiplications, divisions and extractions of roots, doth also cast away from the work itself even the very numbers themselves that are to be

Left: Portrait of John Napier, c. 1600, Scottish mathematician who invented logarithms and mechanical devices for computing.

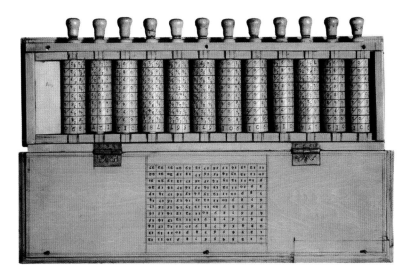

Left: A calculating device created by Scottish mathematician John Napier in 1617, which consists of cylinders inscribed with multiplication tables also known as "Napier's Bones."

multiplied, divided and resolved into roots, and putteth other numbers in their place which perform as much as they can do, only by addition and subtraction, division by two or division by three.

Of course, when he talks about calculators he means a human who is performing a calculation. Electronic calculators were still beyond imagination nearly 400 years ago. And yet, look at a modern calculator and you will see Napier's invention. If there is a "log" button, you are looking at Napier's logarithm. This was his invention for making math easier.

Napier realized that logarithms have some very special properties. They transform difficult operations like multiplication into additions. They transform division into subtractions. They also simplify other tricky operations like roots and powers, turning them into multiplications. In the days when all calculations were performed by hand, these kinds of magical simplifications were as important as the invention of the computer. Suddenly you just needed tables of logarithms and the ability to add or subtract numbers, and you could perform very complicated calculations with speed and accuracy.

Logarithms were considered so important that they were quickly taken up and exploited in many countries around the world. Kepler would never have been able to understand the movement of the planets without logarithms, nor would Newton have been able to understand gravity. All subsequent mathematicians made heavy use of logarithms in their work. Two hundred years later, mathematician Laplace agreed, saying that logarithms:

... by shortening the labours, doubled the life of the astronomer.

Before long, tables of logarithms were optimized onto a gadget known as a slide rule, which enabled the user to slide a little cursor along numbers spaced proportionately to their logarithms, and perform similar tricks to simplify calculations. Slide rules were used until the 1980s when the pocket calculator was developed. Your grandparents, parents (or you) may still own one. Keep it safe.

Logarithms

The logarithm is a difficult-sounding name for something very simple. It comes from looking at basic mathematical operations. When we multiply the same number together several times, we could write 10x10x10x10x10 but it's much easier to write 10^5.

As we saw in an earlier chapter, indices like this can be used to write many types of number:

10^5 = 10x10x10x10x10 = 100,000

10^{-5} = 1 / (10x10x10x10x10) = 0.00001

$10^{1/5}$ = $^5\sqrt{10}$ = 1.58489…

The last result is about 1.58489 because it is the fifth root of 10. Multiply this number by itself five times, and you get 10.

Now try something new. If you take your calculator and type in 1.584893192461114 and press the log key, the result is … 0.2 or ⅕. If you type in 0.00001 and press log, the result is –5. If you type in 100,000 and press log, the result is 5.

In other words, log is simply the inverse of the exponentiation operator:

log $10^{1/5}$ = ⅕

log 10^{-5} = –5

log 10^5 = 5

Pretty easy, right?

If we're using numbers other than 10, then we can specify the number base with the log operator. So:

$\log_{10}$ 100,000 = 5 because as we've seen 100,000 = 10x10x10x10x10 = 10^5

$\log_3$ 81 = 4 because 81 = 3x3x3x3 = 3^4

So the log operator tells us how many times we need to multiply a base number by itself to get the given number. Or in algebra: $\log_a a^b = b$ Logarithms were so useful because of a handy property of indices:

10^3 x 10^4 = 10^{3+4}

Work it out – it's really true. Not only that, but it works for any numbers. So we can say more generally:

10^x x 10^y = 10^{x+y}

Napier's genius was in realizing that the inverse operation of logarithms could be used to turn the multiplication on the left into the addition on the right. In other words:

log (x x y) = log x + log y

So if x x y is rather difficult, we take the log of each number (by looking them up on a table), add the results then look up the result (or the closest result logged) on the log table to find the answer. For example, suppose we wanted to multiply:

2.34 x 3.45

Looking up the logs of 2.34 and 3.45 on our logarithm tables, we find:

0.3692 and 0.5378

We add these (much easier than multiplying the previous numbers) to get:

0.9070

We look this up on our log table and see the closest log is 8.07

So our reasonably accurate calculation looks like this:

2.34 x 3.45 = 8.07

even though we found the answer by adding logs instead of multiplying. Exactly the same trick was used to turn division into subtraction and root into multiplication.

There's one more interesting trick with logarithms that was discovered long after Napier's death. As we saw earlier, log is the inverse of exponentiation and may be applied in different bases:

$$\log_a a^b = b$$

So what happens when we use e as the number base?

$$\log_e e^b = b$$

Or, to write it using the common shorthand (you'll find it on your calculator again):

$$\ln e^b = b$$

Now we're calculating how many times our mysterious value of e should be multiplied by itself. It turns out that the natural logarithm, as it is called, allows us to calculate any other logarithm. To be more precise,

$$\log_b a = \ln a\ /\ \ln b$$

So all that we need to calculate any logarithm table is to create a table of natural logarithms, and divide one value by another. It's another great trick for making calculations easier. It's also just the first of many weird properties of e.

Natural Curves

Jacob Bernoulli (also known as Jacques Bernoulli), the older brother of Johann (who we met in Chapter 0), was born in 1654 in Switzerland. Jacob was the first of the family to go against his father's wishes and study mathematics rather than focus on philosophy and theology. Jacob, like his younger brother, was a very successful mathematician. He was also very argumentative.

Rather than pursue a career in the church, Jacob decided to study and teach mathematics, focusing on the work of great mathematicians before him such as Descartes and Leibnitz. He taught his younger brother Johann mathematics and worked successfully with him, before their relationship was ruined over arguments. According to one writer:

> Sensitivity, irritability, a mutual passion for criticism and an exaggerated need for recognition alienated the brothers, of whom Jacob had the slower but deeper intellect.

One of Jacob's important works was the first discovery of e. It was accidental, for he was interested in finding out what different series of numbers would converge to. He was looking at the idea of compound interest on money. Even in the 17th century, the idea of interest on loans had been well known – this was an important early use for mathematics. Jacob wondered how interest on money could be

Above: Portrait of Jacques Bernoulli, the mathematician who first discovered e.

calculated. He knew that if you added the interest to the sum more frequently (say, monthly instead of yearly) then the sum would increase faster. But what would happen if you calculated it every week? Or every day? Or every second? He quickly discovered that if you deposited $1 at 100 percent APR:

if compounded annually, becomes $2.00
if compounded biannually, becomes $2.25
if compounded quarterly, becomes $2.44
if compounded monthly, becomes $2.61
if compounded weekly, becomes $2.69
if compounded daily, becomes $2.71
if compounded continuously, becomes $2.718.

Right: Euro currency. Bernoulli was interested in the concept of compound interest on money, and the different ways that this could be calculated.

What was this strange new number? What was the significance of 2.718...? We can understand a little more about e (as it became known because of a mathematician called Euler, many years later) by looking at the way in which Jacob investigated the compound interest problem.

This was exciting stuff. Was this weird number another fundamental constant like the golden ratio or pi? It certainly seemed to have something to do with exponentiation for it was created by successive terms raised to a larger power each time. In other words, the first term is to the power of 1, the second term is squared, the third is cubed, and so on.

Bernoulli quickly realized that this new number had something to do with

Bernoulli's compound interest sequence

The problem of compound interest can be written down using a little math. Effectively, Bernoulli wanted to know more about this series of numbers:

$$\left(1+\frac{1}{1}\right)^1 \ \left(1+\frac{1}{2}\right)^2 \ \left(1+\frac{1}{3}\right)^3 \ \left(1+\frac{1}{4}\right)^4 \ \left(1+\frac{1}{5}\right)^5 \dots$$

It seems very simple, but Bernoulli was curious to know what would happen if the sequence was allowed to grow forever. Would the value grow huge, would it shrink to nothing, or would something else happen?

If we work it out, it becomes clear that something else does happen:

1, 2.25, 2.37, 2.44, 2.488, 2.52...

By the time we reach the 100th term in the sequence, its value is 2.704. The further down the sequence we look, the more its value converges to the true value of e. In math we write this as:

$$\lim_{n \to \infty}\left(1+\frac{1}{n}\right)^n = e$$

Which just means: the bigger *n* gets, the closer the value of the equation gets to e.

exponentiation, and logarithms, the inverse of exponentiation. He also realized that it was actually very common in nature. Using e you could construct logarithmic curves, and these spirals seemed to be everywhere: in seashells, flower petals, horns of animals. We saw this spiral in Chapter phi. It's known as the equiangular spiral, but Bernoulli called it the logarithmic spiral, and it's easy to see why.

Bernoulli was so fascinated with the logarithmic spiral that he felt it had almost magical properties. He died aged just 51 years, and following his wishes the logarithmic spiral was carved on his gravestone, with the Latin inscription *Eadem Mutata Resurgo* meaning, "I shall arise the same though changed." Unfortunately the spiral was very crudely carved and was probably not a true logarithmic spiral.

Polar Coordinates

We've seen how Descartes created the Cartesian coordinate system, where x and y give us coordinates along horizontal and vertical axis, and tell us where to draw lines and curves. There's also another system, called Polar Coordinates, that works a little differently. Instead of x and y, we use an angle θ, and a distance r. So in Polar Coordinates, to draw lines and curves we talk about taking an angle and moving a certain distance. It's exactly the same principle used in navigation, where we use a compass to help us find the angle (bearing) and a ruler to figure out how far to walk, sail or fly. It was Newton who first made serious use of Polar Coordinates, but we'll see more of him later.

The best way of drawing a logarithmic spiral is to use Polar Coordinates. Given any angle θ, we can figure out the distance (r) from a center point to draw our curve using this equation (the number b controls how tightly and in which direction the spiral turns):

$$r = a e^{b\theta}$$

Notice e in the middle of the equation? Or, if we know how far we'd like to draw the curve, we can figure out the angle by writing it the other way around (remember that log is the inverse of exponentiation):

$$\theta = 1/b \log_e(r/a)$$

This is why Bernoulli called it the logarithmic spiral.

2.718281828

Pebbles and Fluxions

Isaac Newton was born in 1643 in Lincolnshire, England, 11 years before Jacob Bernoulli. His father died shortly before he was born, so he had a difficult childhood. Newton didn't do particularly well at school, but was very gifted mechanically, making windmills, waterclocks, kites and possibly even inventing some form of human-powered car. But he did not get along with his stepfather and mother and was unhappy at home. He also had quite a temper. Newton wrote at the age of 19 that one of his sins had been:

> Threatening my father and mother Smith to burn them and the house over them.

Eventually he was permitted to study at Trinity College, Cambridge, where he quickly developed a strong interest in mathematics, studying the work of Euclid, Descartes and researching geometry and optics. At the age of 22 Newton received a bachelor's degree from Cambridge, but was forced to leave for two years as the College closed because of the Great Plague of London. Luckily Newton did not fall ill to the Plague, and in 1667 he returned to Trinity College.

When Newton returned, it is likely that two things had already occurred to the young man. One is said to have been the falling of an apple near to him, during his summer in Woolsthorpe.

Above: Isaac Newton, who developed his universal law of gravitation to describe the motion of the planets.

This, combined with his understanding of Kepler's laws, eventually led Newton to his universal law of gravitation several years later. This, at last, was a real explanation for the motion of the planets as described by Kepler. Their motion, like the motion of an apple falling from a tree, was caused by gravity. Newton was persuaded to publish his ideas on gravity in 1685.

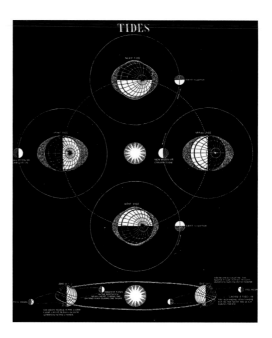

Above: Illustration showing the effects of gravity on tides.

Newton's universal law of gravitation

Newton's universal law of gravitation states that every object in the universe is attracted to every other object, proportionally to the product of their masses, and that attraction is reduced in proportion to the square of the distance between the two objects.

So, given two objects that are a distance r apart, with masses m_1 and m_2: the resulting gravitation force F_g that will move those two objects toward each other is:

$$F_g = G\,\frac{m_1 m_2}{r^2}$$

G is the gravitational constant, which is currently estimated at: 0.000000000066742. In other words, gravity is a very weak force, unless the masses involved are huge (like those of planets and stars). That's why we tend to see apples falling toward the Earth and not the Earth falling toward an apple – the mass of the Earth dominates completely and overwhelms the tiny gravitational pull of the apple. But the Moon is big enough to have a significant effect on the Earth. It orbits the Earth because its mass is smaller, but its pull on the Earth causes a huge bulge in our seas, resulting in our tides.

The second thing that had already occurred to the 23-year-old Newton was something he called fluxions. Newton was thinking about the movement of objects over the "flux" of time. He thought that if we knew where those objects were at many different points in time, then it should be possible to calculate the speed of the objects at any point in time. So if the object was at a point x and y which changed, then the change in x and y would be the fluxions (its instantaneous rate of change), while the changing x and y would be the fluents (the varying or flowing quantity).

If these words seem unfamiliar, it's because they didn't catch on. Unlike his ideas of gravity, the word "fluxion" was not to become part of modern science and mathematics. But Newton's ideas of transforming an equation

*Above: Illustration of
Isaac Newton demonstrating
his spectrum.*

for position into one for speed into one for acceleration was a huge breakthrough. It eventually became known as calculus.

Newton didn't publish his ideas on calculus for years, and that was unfortunate because Leibnitz decided to publish his own work, which turned out to be an independently invented form of calculus. Newton was furious, accusing Leibnitz of stealing ideas during their communication by letter. It was later proved to be nothing more than a misunderstanding caused by delays in postal delivery, but Newton never relented in his attacks of Leibnitz. The resulting feud split mathematicians in England from those in Europe for decades. Unfortunately for Newton, Leibnitz's version of calculus was better developed (we still use his notation today) so the feud set back progress in English mathematics compared to their European counterparts. A young Charles Babbage wrote about the feud nearly two centuries later:

It is a lamentable consideration, that that discovery which has most of any done honour to the genius of man, should nevertheless bring with it a train of reflections so little to the credit of his heart.

Bicycles and calculus

If calculus ever makes you nervous, just remember that it only means "pebble" in Latin. Calculus is essentially all about derivatives of functions and anti-derivatives. To find a derivative, we perform a special operation called *differentiation*. To do the exact opposite (and find the anti-derivative) we perform *integration*.

It sounds complicated, but it's really quite simple. Imagine you're walking up a steep hill. On your Ordnance Survey map you can see the shape of the hill: a big bulge that maybe reaches several hundred feet before dropping down again. Now imagine you've got your bike with you. Although you know the overall shape of the hill, what you *really* need to know is if there are any really steep parts. Any place that's too steep will make it impossible for you to take your bike with you. What do you do? Well, an engineer might dismantle the bike and carry it that way, but a mathematician (like Newton) would try and calculate the steepness of the hill using differentiation. Let's say there's only one way up the hill, and that path is described by the equation:

$$y = 5x^3 - 7x^2 + 3x + 2$$

where y is the height of the hill, and x is the horizontal distance along the ground. So this equation tells us how high the hill is at any given point (it's a bumpy hill). Then, to find the gradient of the hill, we differentiate, which transforms the equation into:

$$y = 15x^2 - 14x + 3$$

This equation tells us how steep the hill is, at any given point. If we plot this and see that it gets too steep anywhere, we know not to take our bike.

Differentiation also allows us to calculate velocity (speed) given details about the position of an object over time, or acceleration, given the speed.

Integration just works the other way around. So if we know the steepness of a hill, we can figure out its shape. (To put it another way: integration allows us to calculate the area under a curve.) Or given the acceleration, we can work out the speed.

There are lots of fiddly rules (that usually have to be learned by heart, which is why many people grow to dislike calculus) about exactly how one equation or function becomes transformed into another. One example is:

In x when differentiated becomes $1/x$

$1/x$ when integrated becomes In $x + c$

(We never quite know if the constant c will exist or not, but we put it there just in case.)

But there is one mathematical expression that is not affected by integration or differentiation. That expression is:

e^x

Our mysterious number e, which so enjoys being used with exponentiation and logarithms, is not changed when differentiating or integrating. Which might seem strange. If our hill was shaped like $y = e^x$ then its gradient would also be e^x. It would be exactly as steep as its height. Or if you were driving and moving along at ex then your speed would be e^x and your acceleration would be e^x. Weird.

Newton's difficult behavior was well known. His assistant and successor Whiston commented:

> Newton was of the most fearful, cautious and suspicious temper that I ever knew.

Nevertheless, Newton's contributions to science and mathematics were substantial.

As well as his work on gravity and calculus, he also correctly understood the properties of light and its behavior through lenses. But less well known are Newton's obsessions with alchemy and theology. He spent more time in laboratories performing experiments than he ever did in mathematics, and he wrote much more on alchemy and theology. Although the Royal Society were given his writings on these topics after his death in 1727, they deemed this work "not fit to be printed." If you'd like to judge for yourself, this is an extract from one of Newton's alchemy manuscripts:

> The spirit of this earth is ye fire in wich Pontanus digests his feculent matter, the blood of infants in which ye Sun and Moon bath themselves, the unclean green Lion which, said Riply, is ye means of joying ye tinctures of Sun and Moon, the broth which Medea poured on ye two serpents, the Venus by medication of which Sun vulgar and the Mercury of 7 eagles saith Philalethes must be decocted ...

In the final years of Newton's life he retired from academic life and took a government job as warden of the mint, responsible for the coinage of England. He became very wealthy in this role, and used his mathematical and chemistry skills to help prevent counterfeiting. Newton spent his final 31 years in this job, and was knighted for his services to his country. He was buried in Westminster Abbey and it is said that after his death, high concentrations of mercury were found in his body — presumably from his many alchemy experiments.

Right: Illustration depicting a meeting of the Royal Society in Crane Court, with Isaac Newton presiding. Undated engraving.

"'ello, 'ello, 'ello, what 'ave we 'ere then? Three blind mice? Playing tick-tack-toe? Now I'm ready, willing and able to give you a chance, but I hear some folk have been caught hook, line and sinker by you rodents. I want you to tell me the truth, the whole truth and nothing but the truth."

THE ETERNAL

CHAPTER 3

There's no red, amber and green light, but I'll say ready-steady-go! Are you ready, willing and able to help? You are! Well three cheers for you. Have you in the past, present or future begged, borrowed or stolen anything? No? Have I got the wrong Tom, Dick and Harry? Well, I'm terribly sorry. Oh, so you're the Tremendously Triumphant Trio of Circus Mice from the triple-ring circus?

A triple back-flip! That's terrific! I can't trump that. OK, I'll leave you alone, but just remember to learn your three R's, your ABCs and eat three good meals a day, and maybe your three wishes will be granted one day. As the French would say, très bon.

Perhaps that bizarre little story didn't make much sense, but it should have illustrated just how often the number three occurs in our lives, speech and even in many words. Try counting the number of times some kind of three appears in the story above.

TRIANGLE

Three is central to many religions: think of the Holy Trinity of Father, Son and Holy Ghost, or the three Nephites. The ancient Babylonians and Celts linked three to creation, for it is a third distinct thing that is born after the union of two. Because of this, three has come to affect our language in many important ways. The word eternity comes from ternity — an obsolete form of trinity. The roots "ter," "tri" and "tre" all derive from three and help form superlatives such as terrific, triumphant and tremendous. Three helps take us from good, to better, to best. And the rhythm of threes proves irresistible in normal speech: "Tom, Dick and Harry," "blood, sweat and tears," "me, myself and I." Our language is filled with threes.

Perhaps not surprisingly, our music is also filled with repeated triplets. Apart from the obvious repetitions: "yeah, yeah, yeah," there are many old and loved songs and rhymes that rely on threes. You may be surprised at the number of these you know (and did you realize they all used three repetitions in a row?):

London Bridge is falling down, falling down, falling down …

Here we go round the mulberry bush, the mulberry bush, the mulberry bush …

Do you know the muffin man, the muffin man, the muffin man?

Mary had a little lamb, little lamb, little lamb …

Polly put the kettle on, the kettle on, the kettle on …

Row, row, row your boat …

The answer is 27 (which is formed by 3³ or 3x3x3 and when added together gives 9, or 3x3).

Above: The triangular pyramid at the Louvre Museum in Paris is constructed out of triangulated glass plates.

There are many superstitions about three. Unlucky events are supposed to occur in threes, but try anything for a third time and you might be lucky. Seeing a three-legged dog is supposed to be lucky, but hearing an owl hoot three times is unlucky. Spitting three times is supposed to shoo away the Devil.

Three is so important to us that we use a triply repeated "truth" to swear people in at court: "Tell the truth, the whole truth and nothing but the truth." We use it to start races — "ready, set, go!" — and we use it to celebrate the winner: "three cheers!"

We have traditionally eaten three meals a day, and our cutlery is a triplet of knife, fork and spoon. If you don't believe three is important, think of three good reasons why not.

Placing Rubber Bands

One shape that exemplifies three perfectly is the triangle. It has three sides (or edges as we call them in mathematics) and three corners (or vertices). Triangles have many important properties, which we'll look at more closely in geometria situs, or "the geometry of place." Today we know it as topology, from the Greek topos, meaning place, and logos meaning study.

In topology we think more about elastic bands than spaghetti. We imagine that we have pushed a collection of push pins into a board, then stretched an elastic band around them. By moving the push pins we change the shape, and topology is the study of how we can make, change, compare and organize shapes formed like this. In fact we normally think of stretching a rubber sheet (perhaps a bit of balloon) over the pins to make a trampoline-like shape rather than just an outline. So in topology there is no great difference between a circle and a square — because one shape can easily be stretched into the other. But there is a big difference between a circle and a figure-eight, for you'd have to cut two holes in the sheet to make the figure-eight.

This is the reason why there's a joke about this form of mathematics:

Q: What do you call someone who drinks from his donut and eats his coffee cup?

A: A topologist.

(I never said it was a funny joke.) In topology, the shape (or "topological space") of a donut is the same as that of a coffee cup. Imagine sticking your thumb in the side of the donut to make a cup-shaped depression and squishing the rest of the ring to make the handle. If you can transform one shape into another by stretching it, but without cutting or filling holes, then the two shapes are topologically equivalent

Chapter π. One property used in computer graphics is their ability to tessellate, or tile perfectly. Throw a pile of uncooked spaghetti onto a flat surface and that surface is now covered with spaghetti triangles. More than three criss-crossing straight lines always form triangles (in Euclidian geometry). If you then make the surface curved (non-planar), and make the triangles small enough you have an approximation of that surface made out of triangles. This is how just about every computer image is drawn: everything is turned into millions of little triangles all carefully placed together, then lighting, color and sometimes photographs are stretched over the result to make it look real.

It turns out that placing shapes is very important in many areas of science and technology. Instead of worrying about the angles or dimensions of the shapes, sometimes it's more important to know where shapes are with respect to each other.

Like any branch of mathematics, this one has a special name. Leibniz, our friend who so loved binary numbers, called it analysus situs, which meant "the analysis of place." It was also called

Right: Mathematician Leonhard Euler who, managed to continue his prolific work even after becoming blind.

(or homeomorphic). This makes more sense when we remember that shapes are nothing more than surfaces (faces) connected together to form edges and vertices (corners). If we stretch a shape, we only change the dimensions of the features, we do not change their number. But cutting a hole in a shape will actually add more edges or vertices (or may completely rearrange the ones you had before). You can easily make a triangle by stretching an elastic band around three drawing pins. If you want a hole in that triangle, you need a new set of drawing pins and another elastic band.

Crossing Bridges

Leonhard Euler was the son of Paul Euler (who had been taught mathematics by Jacob Bernoulli and had even stayed with Johann in Jacob's house while both were undergraduates at university). Leonhard was born in 1707 in Basel, Switzerland. He was educated in a poor school, but his father taught him mathematics and Leonhard quickly excelled in the subject. Before long, Euler was able to convince Johann to tutor him in mathematics. By the age of 16 he had completed a master's degree in philosophy, comparing the ideas of Descartes and Newton. By the time he was 20 he had completed his university studies in mathematics and had already published two articles describing his work.

This was the beginning of what became the most prolific and extraordinary career of any mathematician in history. Euler took a teaching position in the St. Petersburg Academy of Sciences, and at the age of 26 took over the position of senior chair of mathematics from Daniel Bernoulli. He made remarkable advances in number theory (he named many mathematical constants such as e and pi and studied them in depth) and calculus (combining the work of Newton and Leibnitz into the form we know today). He also contributed to practical problems such as cartography, science education, magnetism, fire engines, machines and ship building.

Around this time, Euler studied a problem that helped him create the field of topology. Running through the city of Königsberg, Prussia (now Kaliningrad, Russia), was a large river. The Pregel River divided the city into four distinct regions, separated by water, so seven bridges had been constructed to link the regions together over the river.

Below: Map illustrating the city of Königsberg, Prussia (Kaliningrad, Russia), site of Euler's "Seven Bridges of Königsberg" proof.

KÖNIGSBERG.

Above: Egyptian pyramid, which demonstrates the simple relationship between vertices, edges and faces, found by Euler.

Euler imagined he was walking around the city and posed the question: is it possible to walk over each bridge once and return to the starting point? In 1736 he proved it was not possible (see box opposite).

Euler also was the first to notice a very simple relationship between corners, edges and faces of solid shapes with flat sides (polyhedra). Despite hundreds of mathematicians studying such shapes for thousands of years, it took the insight of Euler to notice that if you add the number of corners (vertices) to the number of faces and subtract the number of edges, the answer is always 2:

$$v + f - e = 2$$

Just to prove he was right, think about an Egyptian pyramid. It has 5 corners, 5 faces (4 triangles and 1 square base) and 8 edges:

$$5 + 5 - 8 = 2$$

Before Euler, mathematicians (such as Pythagorus or Descartes) spent all their time studying the dimensions and angles of shapes. Euler's insight that the relationships between features can be more important than their dimensions enabled him to think in highly original ways.

Königsberg proof

Euler constructed his "Seven Bridges of Königsberg" proof by redrawing the map and making it simpler. Instead of four complicated regions of land and seven bridges, he drew four points and seven links between them.

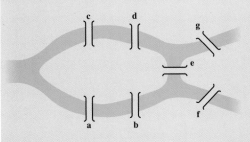

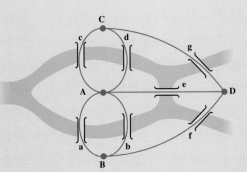

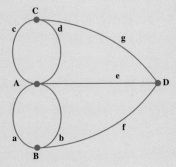

Euler had transformed a complicated shape into a series of nodes and connections — something we now call a graph. He could do this because he understood that the topological properties were important, not the dimensions and distances.

Shrinking the land into blobs (nodes) and stretching the bridges into lines (connections) did not change the topological properties, but it did make the map easier to understand.

For this particular problem, the graph clearly shows that each region of land or node has three connections to the other nodes, and one node has five connections. Euler proved that for any graph, if there is a node with an odd number of connections, it is not possible to traverse the graph following each connection only once. Intuitively, this is quite easy to understand:

If a region has three bridges linked to it, then at some point you will find yourself on the region and unable to leave, or somewhere else and unable to return. You need an even number of connections to be able to get there and back.

A successful traversal of a graph in this way became known as an Eulerian tour or circuit. We still worry about graphs and topologies, and there is a huge amount of research on network topologies, because technologies such as the Internet rely on sensible connections between nodes (computers). Without Euler and his studies of bridges, the Internet might not exist.

Euler was a family man, married at 26 and the father of 13 children, although only five survived to adulthood. Euler claimed some of his best work was performed while holding a baby in his arms, with children playing around him. He suffered from various illnesses, and began to lose the sight in his eyes, possibly because of cataracts. By 1741, when he moved to Berlin and became Director of Mathematics at the new Berlin Academy, he had almost lost the sight in his right eye. Nevertheless his work at the Academy was phenomenal. He supervised the observatory, the botanical gardens, selected personnel, managed finances and the publication of calendars and maps. He advised the government on state lotteries, insurance, pensions and artillery, and supervised work on pumps and pipes for the royal residence of the King. While doing all this, he somehow managed to write 380 articles and several books. He developed new ideas on calculus, planetary orbits, artillery and ballistics, shipbuilding and navigation, the motion of the Moon. He even wrote popular science (much the same as this book). His three books were called *Letters to a Princess of Germany* and comprised several hundred letters that he had written to Princess d'Anhalt Dessau, a niece of Frederick the Great, as he helped educate her from first principles in mathematics and philosophy.

After 25 years in Berlin, at the age of 59, Euler moved back to St. Petersburg. Shortly after his

Above: Photograph of August Möbius, taken c. 1860–65.

return, he fell ill once more and this time lost his sight completely. But Euler continued his mathematics research, now with the assistance of several other mathematicians, including his son. Astonishingly, despite his total blindness, Euler's remarkable memory enabled him to produce almost half his total works (hundreds and hundreds more articles) during this period. Euler contributed to (and helped invent) almost every area of mathematics mentioned in this book. He improved geometry, analytic geometry and trigonometry (we'll see more of this in the next chapter), number theory, calculus, differential calculus, mechanics, acoustics, elasticity, analytic mechanics, lunar theory, the wave theory of light, hydraulics, music and too many other areas to mention. Much of our modern math notation was invented by Euler. In every modern textbook, classroom and laboratory we all use the language of mathematics that Euler helped create.

Wormholes of Paper

Seven years after Euler's death, August Möbius was born, in Schulpforta, Saxony (now Germany). Möbius was educated at home until the age of 13, when he started showing an interest in mathematics. He soon went to the University of Leipzig where he studied astronomy and mathematics (despite his family's wishes that he should study law). He achieved a doctorate by the age of 25 in astronomy and narrowly avoided being drafted into the Prussian army, something he vehemently opposed:

> This is the most horrible idea I have heard of, and anyone who shall venture, dare, hazard, make bold and have the audacity to propose it will not be safe from my dagger.

Möbius began lecturing at the University of Leipzig, and made slow and steady progress as a mathematician. He is remembered best for his work on topology, in particular the deceptively simple shape that bears his name: the Möbius strip. In fact, Möbius did not discover this shape; a mathematician called Listing was the first to think of the idea and publish an article about it. Nevertheless, because of the work he

performed in the area, (perhaps a little unfairly to Listing) it is named after Möbius.

The bizarre properties of the Möbius strip have caused some researchers to liken it to wormholes in space. A wormhole is a theoretical idea that some region of space may be connected to another region — perhaps caused by some kind of black hole — and so traveling through the wormhole would instantly transport

Below: The properties of the Möbius strip have been likened to wormholes in space such as this one.

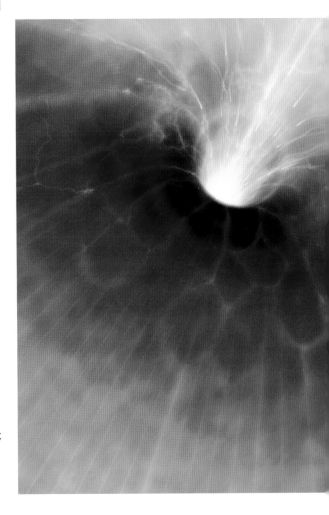

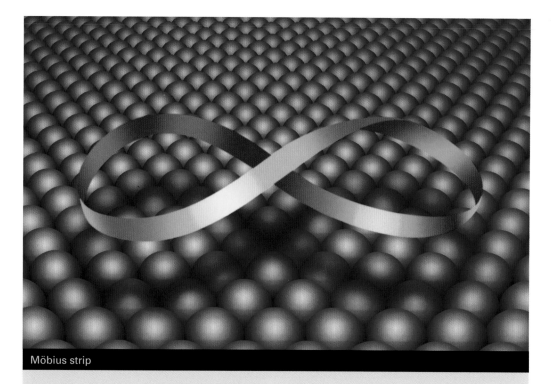

Möbius strip

The Möbius strip is very straightforward to make. Take a piece of paper, and cut a long strip from it. Take both ends, twist one end by 180 degrees and join them together with some tape. You've made a Möbius Strip.

Möbius was fascinated by this strange shape because it has some highly unusual topological features. For example, it has only one side and one edge. If you don't believe this, then make yourself a Möbius strip and run your finger along it, as though you were an ant scurrying along its length. You'll find that you trace all along it, inside and outside until you're back where you started. Follow an edge and the same thing will happen.

One good way to see the weirdness of the Möbius strip for yourself is to cut it lengthways, to try and make two strips. Take some scissors,

carefully make a hole, and cut parallel to the edge, along the center of the strip until you've returned to your starting point. The result is summarized in this little limerick by an anonymous poet:

> A mathematician confided
> That a Möbius band is one-sided,
> And you'll get quite a laugh,
> If you cut one in half,
> For it stays in one piece when divided.

If that made your head hurt, then try this. Make a new Möbius strip, and this time try cutting it into thirds. Again carefully make a hole and cut parallel to the edge, about a third across. You'll quickly discover that although you're trying to cut it into three, you only make one cut. And did you expect the result?

same finding applies to wind. If you imagine winds flowing around the Earth (as they do because of its rotation and the warming of the land and sea masses by the sun), then the same theorem tells us that it is impossible for us to have a steady wind in more or less the same direction, all over the planet. There will always be some regions where winds come together from very different directions, causing cyclones or anticyclones. So even without complications of land and sea, our varied weather is inevitable because of the topology of the Earth.

you to another part of the universe. The Möbius strip can be used to demonstrate this idea. Remember this strange shape has only one side. If we take a hole punch and make a hole through, what have we done? We haven't made a hole from one side to another, because there is only one side! What we've done is make a hole from one region of the shape to another — just like a wormhole.

Today there are many areas and proofs of topology that impact our lives. Topology has grown into a field with many subdisciplines such as combinatorial, geometric, low-dimensional, point-set and (believe it or not) pointless topology. It helps us understand everything from knots to weather. One example of the latter is charmingly known as the Hairy Ball theorem, which states that it is impossible to comb the hair on a hairy ball flat and in the same direction on every part of its surface. Intuitively, you should be able to imagine this: there will always be some part of the ball (perhaps the top and bottom if you comb around the middle) where the hair cannot all be combed flat in roughly the same direction. Using topology we can prove this is so. But perhaps more interestingly, the

Right: Computer-generated artwork representing bent grid lines around an astronomical object, or warped space-time, another example of topology.

Wormholes are thought to be similar to a hole in a Möbius strip. This could therefore allow for jumps through space-time.

Above: Augustus De Morgan, the mathematics professor who first investigated the Map Coloring Problem.

Colorful Maps

On October 23, 1852, Francis Guthrie, a mathematics undergraduate at University College London, asked his professor Augustus De Morgan a question. De Morgan was the first mathematics professor of the new university, and was famous in his own right for his investigations of logic (as we saw in Chapter 2). But despite his own dogged brilliance, De Morgan didn't know

the answer to his student's question. The same day, he wrote to his colleague Hamilton in Dublin:

> A student of mine asked me today to give him a reason for a fact which I did not know was a fact — and do not yet. He says that if a figure be anyhow divided and the compartments differently coloured so that figures with any portion of common boundary line are differently coloured — four colours may be wanted, but not more ...
> If you retort with some very simple case which makes me out a stupid animal, I think I must do as the Sphynx did ...

(De Morgan was referring to the mythological Sphynx who leaped to her death after Oedipus had solved the riddle she had given him. Her riddle was rather easier than the Four Color Conjecture: "What animal walks on four legs in the morning, two at noon, and three in the evening?" The answer is Man (if the day corresponds to a lifetime, then the times correspond to baby, adult and elderly person with a cane.)

But Hamilton didn't know the answer either, replying within three days:

> I am not likely to attempt your quaternion of colour very soon.

The topological problem became known as the Map Coloring Problem, and the question posed by student Francis Guthrie was called the Four Color Conjecture. The problem is most commonly seen when we look at a map of the world. Any map showing all the different countries should color those countries differently from each other so that they are easy to see. The rather obvious rule is that no countries next to each other should be the same color. Francis Guthrie's question was: can you prove that you need no more than four different colors for any map?

While the mathematicians didn't really care whether they helped mapmakers or not, they did find the problem fascinating. For years De Morgan continued asking his colleagues whether they could prove the solution, until eventually a mathematician called Kempe published a proof. Kempe became quite famous, having been elected a Fellow of the Royal Society and knighted. Rather embarrassingly for him, his proof was eventually found to be wrong by another mathematician. Heawood spent 60 years working on the same problem and managed to prove that Kempe was wrong and that five or more colors were needed for any map.

Yet, still the mathematicians were not really satisfied. Five colors could certainly color any map, but what about four? Or even three? The proof took another 80 years and a supercomputer before it could be constructed.

Map coloring

It turns out that you can prove a lot of theorems about map coloring if you feel like it (and mathematicians often do feel like it). Intuitively, a map made from parallel criss-crossing lines only needs two colors — look at a chessboard. Some maps with the right kinds of topological properties only need three colors. But a simple example will show that three is not enough for all maps; here's a "proof by eye":

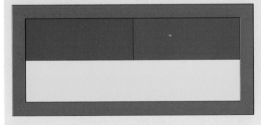

You can't color this with only three colors can you?

The proof of the Four Color Conjecture used a similar method to that used by Euler when he redrew his maps as graphs, where no two nodes linked together should be the same color:

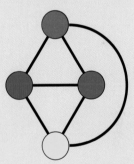

Then a supercomputer was used to calculate many, many special graphs and check that four colors was enough for all of them. The mathematicians had figured out that all other graphs were equivalent to those special ones. The final computer-generated proof was huge and it took years to check it for errors. But to the best of our knowledge, it is true; Francis Guthrie was right. Three might be a magical, special number, but it's not enough to color a map.

Some shapes are more important than others. In our modern world, rectangles and cubes might surround us, but take a walk in the country and you'll see something a little different. With the round sun above you, there may be trees with circular trunks, flowers with circular heads of petals, spherical fruits such as cherries or oranges, and round pebbles: stone washed smooth by water. Drop a pebble in the lake and you'll see circular ripples spreading out. Look at a falling drop of water and observe how it is a perfect little globule. Or blow bubbles and watch the floating wobbly spheres. Turn a circle and look around — and realize you're seeing with circular

A SLICE OF PI

CHAPTER π

eyes. In the natural world, circles are everywhere. But there's only one number that can encompass such a roundness of circularity, and that's pi.

Pi is a ratio. It's the answer to the question: how does the distance from one side of a circle to the other (the diameter) correspond to the distance all the way around the outside (the circumference)? We've known the two measurements of a circle are related for millennia. The challenge was to figure out how. What do you multiply the diameter by to get the circumference? The obvious solution was to draw a circle, measure the circumference and diameter and divide the first by the second. If you try this, you'll find the answer is a little more than 3. So we need to multiply the diameter by a little more than 3 in order to figure out the circumference. But how much more? Exactly what is this number?

Some people, even today, believe that pi equals $\frac{22}{7}$. But this is a rational number (if you look at it in decimal form, it quickly becomes a repeating pattern of numbers). Pi is irrational. Its value is similar to $\frac{22}{7}$, but like all irrational numbers, there is no fraction that can describe

it perfectly, and the decimal expansion will go on forever with no patterns. So like $\sqrt{2}$, phi and e, pi is a natural constant that is impossible to know completely. That hasn't stopped many from trying.

Making a Circle

Ironically, the first investigations of pi were made before zero had even been invented. We've already met one of the first and most impressive pi-seekers: Archimedes (who if you recall was born in Sicily 300 years after Pythagoras and just before Euclid died). As well as inventing levers, pulleys, ship-crushing devices and spiral pumps, Archimedes spent much of his time worrying about circles and spheres. He wrote several books (or scrolls, as they were in those days) on the subject: *On the Sphere and Cylinder, On Spirals, On Conoids and Spheroids* and *Measurement of a Circle*.

It is said that he used his skills in all things spherical to make two spheres, later taken by the invading Roman General Marcellus. One was a celestial globe with the stars and constellations painted or engraved on it. The other was a working planetarium, showing the circular motion of the Sun, moon and planets as seen from the Earth. To say that the Romans were impressed would be a big understatement. In the words of one such Roman called Cicero,

Archimedes must have been "endowed with greater genius that one would imagine it possible for a human being to possess" to be able to build such an unprecedented device.

Archimedes was the first to realize that pi was irrational, and that its value was not $^{22}/_7$. He knew he might never figure out the exact value of pi, so he used the following trick to work out approximately what it was.

Approximating pi by trapping circles

Archimedes decided to use polygons (shapes made from straight sides) to approximate circles. He drew one on the outside of the circle and one within the circle and then figured out what the ratio of perimeter (circumference) was to the diameter for both polygons. Because the outside shape was bigger and the inside one smaller, he knew that the true value of pi lay between the two ratios.

As a simple example, imagine he used just four-sided polygons (squares). If the side of the largest square is D, then its perimeter must be 4D, and obviously the diameter from side-to-side is also D. The first ratio is thus 4D/D, which is 4.

You can work out that the perimeter of the smaller square is 4D/√2, and this time measuring corner to corner, the diameter is still D. So the second ratio is 4D / √2 / D = 4 / √2 = 2.828427...

So we know that pi is smaller than 4 and bigger than 2.828427...

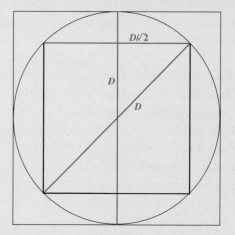

Now all we do is repeat the same idea but use polygons with more sides, so that they approximate the shape of the circle better. Archimedes used polygons that had 96 sides to show that pi lay between $3^{10}/_{70}$ and $3^{10}/_{71}$. Or to write that in a more simplified form, between: $^{22}/_7$ and $^{223}/_{71}$.

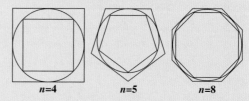

$n=4$ $n=5$ $n=8$

Archimedes' trick was ingenious, using polygons to "trap" the circle and find pi. His method was so accurate that no one found a better value for 500 years. Because he did not try to find a decimal expansion (unlike many mathematicians who came after him), Archimedes will always be right. Pi will always lie between $\frac{22}{7}$ and $\frac{223}{71}$, no matter how many decimal places we expand it to.

Although Archimedes discovered pi to a remarkable accuracy and created some extraordinary inventions (his nicknames were "the wise one," "the master," and "the great geometer") he considered one mathematical finding to be his greatest achievement. Once again, it was related to pi, but this time it was about the volume of a sphere and cylinder. He managed to prove that if you took a sphere and wrapped a cylinder around it so that its sides touched the edges of the sphere, then the volume of the sphere is two-thirds the volume of the cylinder. He also proved that the ratio of area to volume of the sphere is the same as the ratio of area to volume of the cylinder. These findings were a huge breakthrough in an era when people previously had no idea how to calculate volumes and surface areas of shapes such as spheres.

Above: Engraving (c.46–c.120) of Plutarch, the Roman writer who recorded information about Archimedes' life.

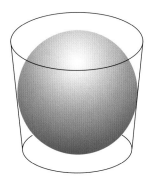

Above: A sphere wrapped in a cylinder has a volume two-thirds that of the cylinder.

While we only know of Archimedes through his work (the only biography written about him was lost millennia ago), quite a bit has been written about his death, which occured in 212 B.C. at the age of 75. Three versions of his death were reported by the Roman writer Plutarch. Be a detective and see which you think really happened:

Version 1: "Archimedes was, as fate would have it, intent upon working out some problem by a diagram, and having fixed his mind alike and his eyes upon the subject of his speculation, he never noticed the incursion of the Romans, nor that the city was taken. Lost in study and contemplation, a soldier, unexpectedly coming up to him, commanded him to follow to Marcellus.

He declined, before he had worked out his problem to a demonstration, so the soldier, enraged, drew his sword and ran him through."

Version 2: "A Roman soldier, running upon him with a drawn sword, threatened to kill him; and Archimedes, looking back, earnestly besought him to hold his hand a little while, that he might not leave what he was then at work upon inconclusive and imperfect; but the soldier, nothing moved by his entreaty, instantly killed him."

Version 3: "As Archimedes was carrying to Marcellus mathematical instruments, dials, spheres and angles, by which the magnitude of the sun might be measured to the sight, some soldiers seeing him, and thinking that he carried gold in a vessel, slew him."

Above: Illustration describing one version of the last moments of Archimedes' life, in which he was interrupted from his work by a Roman solider.

The first two versions seem remarkably similar, and a popular myth has grown around them, whereby Archimedes who, when interrupted by a Roman soldier while drawing figures in the sand, utters his final words as: "Don't disturb my circles!"

We'll never know what actually happened on that day. We do know that the invading General Marcellus was furious with the soldier, for he had ordered that the brilliant Archimedes be kept alive. Marcellus had the soldier executed for disobeying his orders.

Despite the successful Roman invasion, Archimedes was honored in death, and his wishes for the design of his tombstone were respected. On it was carved a cylinder containing a sphere, with his theorem describing that the volume of the sphere is two-thirds that of the cylinder, and some say, an inscription of pi. His tomb survived for at least a hundred years. In 75 B.C. the Roman statesman Cicero wrote about the following adventure:

When I was quaestor in Sicily I managed to track down Archimedes' grave. The Syracusans knew nothing about it, and indeed denied that any such thing existed. But there it was, completely surrounded and hidden by bushes of brambles and thorns.

Below: Painting depicting Cicero's discovery of Archimedes' grave.

I remembered having heard of some simple lines of verse which had been inscribed on his tomb, referring to a sphere and cylinder modelled in stone on top of the grave. And so I took a good look round all the numerous tombs that stand beside the Agrigentine Gate.

Finally I noticed a little column, just visible above the scrub: it was surmounted by a sphere and a cylinder. I immediately said to the Syracusans, some of whose leading citizens were with me at the time, that I believed this was the very object I had been looking for. Men were sent in with sickles to clear the site, and when a path to the monument had been opened we walked right up to it. And the verses were still visible, though approximately the second half of each line had been worn away.

So one of the most famous cities in the Greek world, and in former days a great center of learning as well, would have

remained in total ignorance of the tomb of the most brilliant citizen it had ever produced, had a man from Arpinum not come and pointed it out!

Thankfully, the lack of interest in mathematics shown in Roman times was short-lived, and today Archimedes is gone but not forgotten.

How to Make a Pi

Pi continued to fascinate mathematicians for centuries, and slowly we came to know it more accurately. Many of the pioneers of numbers that we met in earlier chapters were involved in these studies. For example, Brahmagupta (who helped understand zero) thought that pi was equal to $\sqrt{10}$. He was wrong; they are only

the same to a single decimal place. One hundred and sixty years later, our friend Al-Khwarizmi (who invented *al-jabr*) managed to figure out the value of pi to four decimal places, writing that it was: 3.1416. This was better than the attempt of Fibonacci (remember his rabbits?), some 400 years later, who used another 96-sided polygon to estimate the value of pi as $^{864}/_{275}$.

The mathematicians continued to use variations based on Archimedes' method for centuries. One of the most impressive was made by a German mathematician in 1596. Van Ceulen spent most of his life calculating the value of pi by using a polygon with an almost unbelievable 4,611,686,018,427,387,904 sides. He figured out pi to 35 decimal places:

3.14159265358979323846264338327950029

The result of his astonishing labors was engraved on his tombstone when he died aged 70.

Around this time, mathematicians began to notice that pi could be found using easier methods than polygons with horrifying numbers of

Calculating pi

In the 17th century, a mathematician named Wallis discovered this strange series that produces a multiple of pi:

$2/\pi = (1 \times 3 \times 3 \times 5 \times 5 \times 7 \times \dots) / (2 \times 2 \times 4 \times 4 \times 6 \times 6 \times \dots)$

And at around the same time a mathematician named James Gregory discovered this famous series (although sometimes Liebnitz is credited for its discovery):

$\pi / 4 = 1 - \frac{1}{3} + \frac{1}{5} - \frac{1}{7} + \dots$

Neither are very useful for calculating the value of pi, for you need tens of thousands of terms in the series before the value starts becoming close to the true value of pi. But Gregory also calculated a more useful series that converges more quickly:

$\pi / 6 = (1/\sqrt{3}) (1 - 1/(3 \times 3) + 1/(5 \times 3 \times 3) - 1/(7 \times 3 \times 3 \times 3) + \dots$

You only need nine terms in this series before the result gives pi correctly to four decimal places.

*Right: Georges Louis Leclerc,
Count of Baffon, or George
Buffon, who invented Buffon's
needle problem.*

sides. Despite the fact that pi is inextricably
linked to circles, multiples of pi magically
appeared out of some sequences of numbers.

　Using these methods and other similar ones,
mathematicians over the following centuries
calculated more and more of pi. Perhaps one of
the more bizarre methods was proposed by
French scientist Georges Buffon in the 18th
century. Buffon calculated that if a needle is
dropped on a tiled floor, then the probability that

the needle lies across the edge of a tile is $2k/\pi$
(where k is the length of the needle and its
value is less than 1). In 1901, a man called
Lazzerini used the idea to calculate the value of
pi. He dropped his needle 34,080 times,

Above: Portrait of Augustus De Morgan, the mathematician who discovered the flaw in William Shanks' value of pi.

counted the occasions where the needle landed across a tile edge and calculated pi correctly to six decimal places. But other mathematicians were suspicious, pointing out that if you already knew the value of pi, you just needed to stop your needle-dropping experiment at the right time to ensure an accurate result. By dropping his needle 34,080 times (and not a round number such as 30,000 or 35,000 times), Lazzerini had cheated and ensured his answer was more accurate than it would have been otherwise. In a cheeky paper that tried to show how such cheating could be done, a mathematician called Gridgeman demonstrated that by throwing a needle of length exactly 0.7857 just twice, and having it hit the edge of a tile once, the formula resulted in a good approximation of pi, because:

$$2 \times 0.7857 / \pi = \frac{1}{2}$$

so $\pi = 3.1428$

Although he was poking fun at the previous experimental attempts at calculating pi by dropping needles, Gridgeman was making a serious point: if you already know the value of pi to a certain accuracy, it is easy to fake an experiment that appears to create pi to the same (or worse) accuracy.

With the advent of easier ways to calculate pi (using series of numbers rather than needles), mathematicians were soon calculating this popular number to hundreds of decimal places. By 1874, a mathematician called Shanks had found pi to an amazing 707 decimal places. And yet something was amiss. De Morgan (the same mathematician who had been so puzzled by the four-color conjecture in the previous chapter) noticed something rather strange about the numbers in the value of pi found by Shanks. More specifically, De Morgan noticed that after about the first 500 decimal places, there didn't seem to be nearly as many 7s.

This was a weird mystery, for it was well known that the distribution of numbers in the decimal expansion of pi always seemed very even: there always seemed to be roughly the same numbers of each digit 0 to 9 in the value of pi. It was one of the astonishing things about pi. Tossing a 10 sided die would produce the same frequency of digits 0 to 9 in a sequence that has no patterns (because the digits are in a random order). It was always believed that pi also contained the same number of digits from 0 to 9 with no patterns, but with digits far from random. It's a weird paradox of this irrational number: a non-random patternless number with the properties of a random number except that each of its digits are always the same.

So De Morgan finding such an inconsistency in pi — the lack of 7s — was very strange. The mystery remained until 1945 when another mathematician called Ferguson calculated pi to 620 decimal places and realized that Shanks had made a mistake. All the digits after the 528th place were wrong in Shanks's calculation. In the correct version, there was no inconsistency: each digit 0 to 9 appears in pi with equal frequency.

After 1947, desk calculators and computers were used by mathematicians to calculate ever—more digits of pi. For many years, the power of a new computer was demonstrated by showing how many digits it could calculate. By 1999, a Hitachi SR8000 supercomputer had calculated 206,158,430,000 digits of pi. That's nearly six billion times more digits than Van Ceulen found after spending his life on the task. Today even a desktop computer is so powerful that it could easily calculate trillions of digits of pi in seconds, so the fascination of this strange number is fading. Which is a shame, for of course there are still an infinite number of digits in pi left to find.

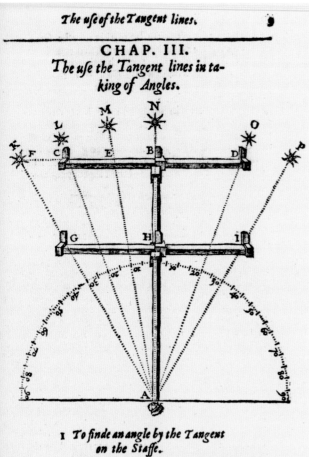

The use of the Tangent lines. **9**

CHAP. III.
The use the Tangent lines in taking of Angles.

1 *To finde an angle by the Tangent on the Staffe.*

LEt the midle sight be alwaies set to the middle of the Crosse, noted with 20 and 30, and then the Crosse
B b drawne

Measuring Angles

In the previous chapter we saw how triangles with their three sides and three corners are fundamental to geometry and topology. But triangles also have three angles inside each corner — a fact so important to mathematicians over the ages that yet another type of mathematics was invented to describe them. From the Greek *trigonon* (meaning three angles) and *metro* (meaning measure) came trigonometry.

Trigonometry is the mathematics of angles. If you have two straight lines that meet at a point, then they must be at some angle to each other. The oldest and most common way of measuring angles is in degrees, with 360 degrees in a full circle. The easiest way to think about this is to look at the two hands of a clock. When both hands point in the same direction, say 12 o'clock, it's as though we have two lines on top of each other, so the angle between them is 0 degrees. With one hand at 12 and one at 9, the angle between them is 90 degrees. With one at 12 and one at 6, the angle is 180 degrees, and with one at 12 and the other at 3, the angle is 270 degrees (we normally measure angles counterclockwise). Angles are very important when thinking about geometric shapes such as triangles, squares and pentagrams. We often use the smaller angle or the internal angle of geometric shapes — it makes more sense to talk about the internal angle of a square being 90 degrees rather than 360 − 90 = 270 degrees.

The reason why there are 360 degrees in a full circle is almost certainly because of the Babylonians living before 300 B.C. These ancient people counted using base 60 (in contrast to our base 10, or our computer's binary base 2). They probably divided the circle into 60, then divided each of those segments into 60, resulting in our

Pl. 15.

Fig. 176.

Fig. 177.

Fig. 180.

Fig. 178.

Fig. 179

Fig. 181.

Astronomie, Quart de Cercle Mobile.

Benard direx.

10

360 degrees. When navigating and describing the position of the stars, they measured degrees counterclockwise around the Earth.

Trigonometry arose out of a simple problem: if you have a triangle and know only some of the measurements of it, how do you calculate the other measurements? For example, if you know two angles and the length of just one side, how can you calculate the length of the other sides?

One partial solution to this problem for several centuries was the chord, which was thought up by an astronomer called Hipparchus. Born in 190 B.C., in what was then Nicaea, Bithynia (and is now Iznik in northwest Turkey), most knowledge of Hipparchus has been lost over the centuries. We know next to nothing about the man who was to become a highly successful Greek astronomer. But we do know a little about his work. During a solar eclipse Hipparchus measured the different parts of the Moon visible at different locations, and from this he was able to calculate how far the Moon was from the Earth. He was astonishingly accurate, estimating the distance to be between

The origin of the chord

This turns out to be a crucial question when surveying land. Imagine you are on one side of a river and you want to draw an accurate map and figure out exactly how far away an important feature is, say a church. You can move to two positions A and B on this side of the river. What do you do?

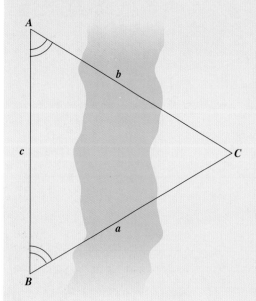

Well, if we say the church is at position C, then we've made a huge triangle on the ground. We can measure the distance from A to B, and using a little viewing device (a surveyor's telescope, or theodolite) we can measure the angles to point C (we measure the angle between AB and BC, and between AB and AC). So we now know the length of one side of the triangle, and two of its internal angles. Now we need to know how far point C is from A or from B. How do we find the lengths of the other two sides of the triangle? We can't use the Pythagorean theorem, because it only works for right-angle triangles (which must have one angle equal to 90 degrees). We also would need to know the length of two sides to calculate the third, and we only know the length of one. Somehow we need to use the angle between sides to help tell us the length of a side.

Above: Illustration showing the Greek astronomer Hipparchus.

59 and 67 Earth radii — today we know the true distance is 60. He also calculated the length of the Earth's year with such accuracy that he discovered and measured the precession of the Earth (the movement of the Earth's axis of rotation over time). Quite an achievement for someone living over 2,000 years ago!

Hipparchus was also the first person to create a table of chords — a list of triangle side lengths that corresponded to different angles of a triangle. This was the beginning of trigonometry.

One of the most important astronomers and geographers was born around the year A.D. 85 in Egypt. Claudius Ptolemy believed in Aristotle's geocentric view of the heavens, in which it was believed that everything orbited the Earth. He was unusual in that much of his mathematics was developed to explain real observations of the movement of planets, but because much of the information he used was inaccurate, his theories were not proved wrong for centuries. Later scientists criticized Ptolemy, accusing him of faking his evidence to support his incorrect ideas. Newton was especially nasty, saying that Ptolemy:

Left: Sixteenth-century print
of Claudius Ptolemy, from a
book published by Nicolo
Bascarini in 1548.

Chords

Given a circle with centre O, and two points on
the circle A and B, the chord AOB is simply the
length of the line AB. Varying the angle at O
varies the length of the chord from zero (when
the angle is zero) to the diameter of the circle
(when the angle is 180 degrees). Hipparchus
simply produced a table of results for different
angles, knowing that the relationship between
angle and length would be the same, no
matter how big the circle. So to figure out the
true length of a chord, he just figured out how
big the circle should be, then scaled the result
from his table by the same amount.
Hipparchus made a table of chords for angles
7.5 degrees apart (a total of 24 in 180 degrees).

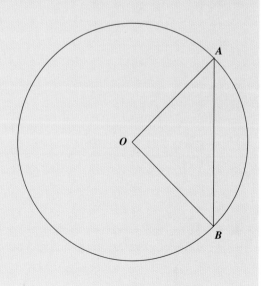

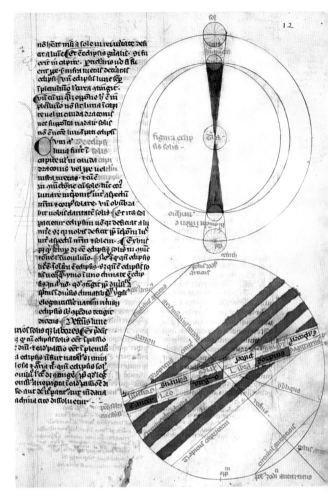

developed certain astronomical theories and discovered that they were not consistent with observation. Instead of abandoning the theories, he deliberately fabricated observations from the theories so that he could claim that the observations prove the validity of his theories. In every scientific or scholarly setting known, this practice is called fraud, and it is a crime against science and scholarship.

But such comments are probably unfair, when it is recalled that Ptolemy lived 2,000 years ago at a time when scientific method was still young and we had little ability to obtain accurate measurements of the movement of planets.

Despite the forgivable flaws in some of his ideas, the mathematics that Ptolemy developed was enormously important. He wrote many books, including a 13-volume epic that became known as the *Almagest*, about the movement of the planets. His ideas (about circular orbits around the Earth) were not superseded for another 1,400 years (by people such as Kepler), and so his books were considered by many to be as important as Euclid's *Elements*. In the course of his work, Ptolemy calculated pi correctly to four decimal places, which at the time was the most accurate figure ever found and was not improved on for another 150 years. He also developed the idea of chords further, producing a table of chord values for angles half a degree apart, and figured out many clever rules and operations that allowed chords to be manipulated. Through this work, he

Above: Mediaeval artwork showing a solar eclipse. The top diagram shows the Sun (yellow) and Moon (partly green) orbiting the Earth.

Trigonometry

While chords were very intuitive, they were not the most useful when trying to figure out the missing sides of a triangle (as we need to for land surveying). The sine function (or "sin" on your calculator) was the solution. Instead of telling you the length of a line opposite the angle (between two points on a circle) the sine function tells you the length of a line from the center of the circle to the edge for a given angle. To calculate the sine of an angle x, we measure the angle x counterclockwise, draw a line from the center to the edge of a circle with radius 1 (point P), then the y-coordinate of P is the answer. That's why sine of 0 degrees is 0 and sine of 90 degrees is 1.

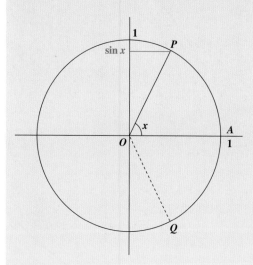

Another way of looking at sine is to consider it as half a chord. If we were to reflect the line OP in the x-axis (put a mirror along the line OA) and create a new point Q, then PQ is a chord. This is the origin of sine, and we know this because we know the history of the word itself. The Sanskrit word for chord-half was *jya-ardha*, which was sometimes shortened to *jiva*.

In Arabic this became *jiba*, written simply as *jb*. Latin translators mistook *jb* for the Arabic word *jaib*, which meant breast, and so used the Latin word *sinus*. In English, *sinus* became sine.

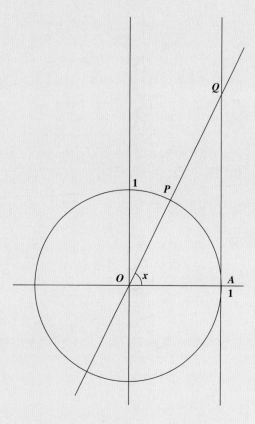

Once you have sine, it's easy to think about similar functions. Cosine, for example, can be calculated by reading the x-coordinate of point P instead of the y-coordinate. That's why cosine of 0 degrees is 1 and cosine of 90 degrees is 0. There's also a related function called tangent which works exactly the same as sine, except that instead of worrying about point P which sits on the circle, we worry about a point Q, that sits on a line *AQ* at a tangent to the circle that intersects with the continuation of the line *OP*.

Once again, reading the y-coordinate of point Q gives the tangent of angle *x*. That's why the tangent of 0 degrees is 0 and the tangent of 90 degrees is undefined (we don't know where Q will be when *x* is 90 degrees, so the answer is undefined, not infinity). This is one calculation that some electronic calculators actually get wrong. See what yours says for tan 90.

All of the trigonometry functions can also operate backward, so to go from the length of the line back to the angle, we just use the three inverse functions: arcsine, arccosine, arctan (which are sometimes written: sin^{-1}, cos^{-1}, tan^{-1}, or asin, acos, atan).

Because all trigonometry functions are related to the same triangles and circles, there are many rules and formulae that help us to use them. Unless you're a mathematician they're not very interesting, especially when you have to learn them by heart, but they can be very useful. Here's just three out of many tens of examples:

$$\sin^2 A + \cos^2 A = 1$$

$$\sin(A+B) = \sin A \cos B + \cos A \sin B$$

$$\sin 2A = 2 \sin A \cos A$$

laid the foundations for the next generation of trigonometry functions, which became known as sine, cosine and tangent.

Surfing Sine Waves

You don't need a sine function to have a sine wave. It turns out that sine waves are really common things. Light moves as though it's a sine wave, with different colors corresponding to different frequencies. The waves are measured in "wavelengths," which rather obviously means "length of the wave" (so a short wavelength is a sinewave squashed closer together like a compressed spring and a long wavelength is a sine wave stretched further apart). Red has a longer wavelength than green, and green has a longer wavelength than blue. But visible light (the wavelength our eyes can see) is just a tiny,

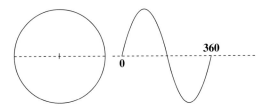

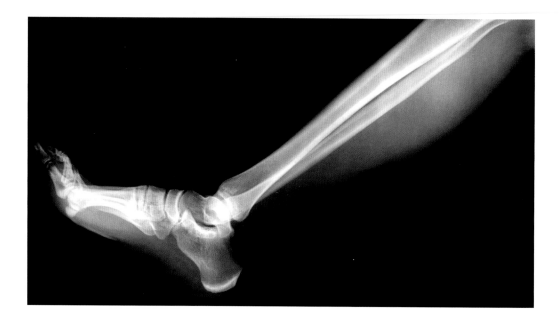

Above: X-ray of a human foot.

tiny part of the electromagnetic spectrum (and all of the wavelengths thrown out by the Sun, for example). Your infrared remote control acts like a torch, shining pulses of light with a larger wavelength than red. Your microwave oven has an even larger wavelength, and radio waves have wavelengths that are bigger still. And we use plenty of wavelengths that are smaller than those in the visible spectrum too. Ultraviolet, or UV light, has a wavelength smaller than violet. X-rays, used in the hospital because everything except our bones is transparent to this kind of light, have a wavelength that is even smaller. Gamma rays are even smaller. In the Marvel comics, the Incredible Hulk was supposed to be the result of an accident involving gamma rays — but in reality they form a type of radiation sometimes used in medicine to sterilize instruments or in radiation therapy.

Light, or electromagnetic radiation, is not the only type of sine wave out there. All your alternating current (AC) electricity alternates because it is supplied as a sine wave of electricity, with a frequency of around 50 or 60 hertz, depending on which country you live in. If the electricity was water, it would act rather as if the water was being pushed, then pulled, then pushed, then pulled through your water pipes in a sine wave of pressure. Electricity is supplied like this because it's a more effective way of transferring power down long cables from the power stations to your home, compared to simply trying to push it all the time using direct current (DC), as batteries supply.

Just as all the colors of light are made from sine waves of different wavelengths (frequencies), so sound is made from vibrations of different frequencies. Put your hand in front

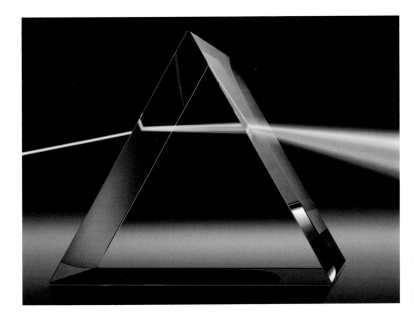

Left: Light beams shown through a glass prism. Red, green and blue wavelengths are all visible.

Right: Electromagnetic spectrum (center), its colors of visible light (bottom) and changing wavelengths of electromagnetic radiation (top).

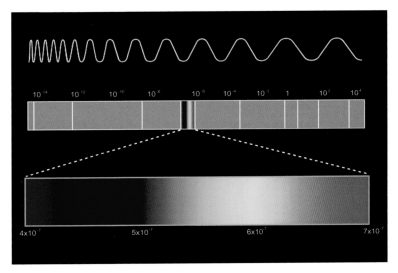

of a loudspeaker and you may feel the pressure of the sound waves; or stand close to a deep booming drum and you'll feel the sound waves vibrate inside your chest. Loudspeakers work by vibrating a cone of cardboard at different frequencies, pushing pulses of vibration to your ears. Feed a loudspeaker a sine wave and you'll hear a perfect tone in a single frequency. Add lots and lots of sine waves of different

frequencies together and you can make any noise possible, from a symphony to the sound of your own voice. Frequencies are very important in music (as was understood by Pythagoras over 2,000 years ago). Our musical notes are divided into octaves, with 12 notes (or semitones) in each. An octave is simply a doubling of the frequency, so the note A has a frequency of 220 Hz, and one octave higher, A1 has a frequency of 440 Hz. This is why some notes sound harmonious when played together: their sine waves line up nicely (sine waves of 220 and 440 Hz spend much of the time pushing vibrations of sound at the same time). Discordant notes sound bad because their frequencies do not overlap; producing a messy and uneven pile of sine waves.

Pendulums and Heresy

Sine waves and pi are also to be found in other places. Imagine the pendulum of a clock, swinging back and forth. Plot its side-to-side motion over time and you have another sine wave. It was an Italian man named Galileo Galilei, born seven years before Kepler in 1564, who was to make the most important breakthroughs on pendulums.

Galileo's father was a teacher of music who had performed experiments on strings to understand frequencies and harmony. But his father wanted Galileo to become a medical

Below: Portrait engraving of Galileo Galilei explaining his theories to a skeptical monk.

Right: Two balls, one made from wood and one from metal are dropped from the leaning tower of Pisa, and were "seen to fall evenly." Galileo probably never performed this experiment in real life.

doctor, and so after having him educated in a monastery, sent him to the University of Pisa to earn a medical degree. Galileo was not very interested in medicine and spent more time studying mathematics and natural philosophy. Despite his best efforts to persuade his father (even taking one of his mathematics professors to meet his stubborn dad), Galileo was forced to continue his medical degree, studying mathematics in his spare time. But eventually, by

the age of 21, Galileo gave up the medical degree, becoming a private mathematics tutor instead. When he was 25, Galileo managed to secure a job as a professor of mathematics at the University of Pisa and a few years later managed to secure a much better-paid job as a mathematics professor at the University of Padua.

Much of Galileo's work was about the motion of the planets and other falling objects. This led to the popular story that Galileo had performed

Above: Galileo performed many experiments using pendulums, and emphasized the importance of the length of the pendulum.

a famous experiment, dropping a ball made from wood and one from metal from the top of the Leaning Tower of Pisa, and observing that both hit the ground at the same time. It's unlikely that Galileo ever bothered to perform the experiment (although we know that other mathematicians such as Simon Stevin did have a go). But we do know that Galileo understood the principle very well, for he had described his revelation during a hailstorm. He noticed that both small and large hailstones hit the ground together, so unless all the large hailstones were falling from much higher up (which seemed very unlikely), this meant that objects must fall at the same rate, regardless of their weight.

Galileo performed many experiments with pendulums, discovering that the period of a pendulum (the time it takes to swing back and forth) does not depend on the weight of the swinging bob, nor does it depend on how far it swings. But it does depend on how long the pendulum is: make the pendulum four times as long, and it will take twice as long to swing back and forth. These were all important discoveries that contradicted the Aristotelian view of the universe (that heavier objects fell faster and would therefore swing quicker on a pendulum).

But Galileo was only just beginning. By 1609 he had heard of a magical "spyglass" that somehow made objects look bigger. He wrote about this discovery in 1610:

About ten months ago a report reached my ears that a certain Fleming had constructed a spyglass by means of which visible objects, though very distant from the eye of the observer, were distinctly seen as if nearby. Of this truly remarkable effect several experiences were related, to which some persons believed while other denied them. A few days later the report was confirmed by a letter I received from a Frenchman in Paris, Jacques Badovere, which caused me to apply myself wholeheartedly to investigate

means by which I might arrive at the invention of a similar instrument. This I did soon afterwards, my basis being the doctrine of refraction.

Below: Galileo's telescopes were among the world's best during his time, with magnifications of 8x and 9x.

Using remarkable skill and ingenuity, Galileo figured out how to grind and polish glass into lenses and proceeded to make the world's best telescopes, with magnifications of 8x or 9x. This was a breakthrough that had some obvious military applications — you would be able to see enemy ships long before they could see you. But when Galileo pointed his new telescope to the heavens, the universe was changed forever. His were the first human eyes to see mountains on the Moon, the stars of the Milky Way and even some of the moons of Jupiter. According to one professor:

> In about two months, December and January, he made more discoveries that changed the world than anyone has ever made before or since.

Just weeks after completing his telescope, Galileo published a book called the *Starry Messenger*, describing his amazing new discoveries. He became a celebrity overnight and quickly took a new position as Chief Mathematician at the University of Pisa (without any teaching duties) and "Mathematician and Philosopher" to the

Right: Illustration of Galileo as he demonstrates his work to the Venetian Doge and all of his senators.

Grand Duke of Tuscany. Before long, he noticed that Saturn appeared to have strange "ears" sticking out from its sides (his telescope was not quite good enough to resolve Saturn's rings). He discovered sunspots and he also noticed that Venus showed phases, like the phases of the Moon — an important fact that strongly suggested that Venus was orbiting the Sun and not the Earth.

By 1616 Galileo was convinced that the old geocentric view, that the Earth was the center of the universe with everything orbiting around it, was wrong. Despite the longheld teachings of Aristotle and Ptolemy, Galileo thought he knew better. He wrote in a letter:

I hold that the Sun is located at the centre of the revolutions of the heavenly orbs and does not change place, and that the Earth rotates on itself and moves around it. Moreover ... I confirm this view not only by refuting Ptolemy's and Aristotle's arguments, but also by producing many for the other side, especially some pertaining to physical effects whose causes perhaps cannot be determined in any other way, and other astronomical discoveries; these discoveries clearly confute the Ptolemaic system, and they agree admirably with this other position and confirm it.

Galileo and Kepler were two isolated voices among the orthodox crowd, and Kepler was too nervous to publish his views. Galileo was also to write:

> I wish, my dear Kepler, that we could have a good laugh together at the extraordinary stupidity of the mob. What do you think of the foremost philosophers of this University? In spite of my oft-repeated efforts and invitations, they have refused, with the obstinacy of a glutted adder, to look at the planets or Moon or my telescope.

The religious world was not ready for Galileo's revelation. Pope Paul V ordered the cardinals of the Inquisition to investigate the matter. An official religious truth was declared: the teachings of Copernicus were condemned; the Earth was the center of the universe. But soon a new pope was elected: Pope Urban VIII. He appeared to be more receptive to Galileo's ideas, encouraging Galileo to write them down. After six years of effort, Galileo published his findings in a book called *Dialogue Concerning the Two Chief Systems of the World — Ptolemaic and Copernican*. Shortly after its publication, the Inquisition banned its sale and Galileo was found guilty of heresy and sentenced

to lifelong imprisonment — which meant he was under house arrest for the rest of his life.

Despite being watched by officers of the Inquisition, Galileo continued his work. Two years before he died, he realized that the regular swinging of pendulums could be used in clocks, but he never saw his idea used in practice. Galileo died at the age of 78. It took another 350 years for the Catholic Church to admit that "mistakes had been made in the case of Galileo" in a statement made by Pope John Paul II in 1992.

Below: The 1632 title page of Galileo's Dialogo, *or* Dialogue Concerning the Two Chief Systems of the World.

DIALOGO
D I
GALILEO GALILEI LINCEO
MATEMATICO SOPRAORDINARIO
DELLO STVDIO DI PISA.
E Filosofo, e Matematico primario del
SERENISSIMO
GR.DVCA DI TOSCANA.
Doue ne i congressi di quattro giornate si discorre
sopra i due
MASSIMI SISTEMI DEL MONDO
TOLEMAICO, E COPERNICANO;
*Proponendo indeterminatamente le ragioni Filosofiche, e Naturali
tanto per l'vna, quanto per l'altra parte.*

CON PRI   VILEGI.

IN FIORENZA, Per Gio:Batista Landini MDCXXXII.

CON LICENZA DE' SVPERIORI.

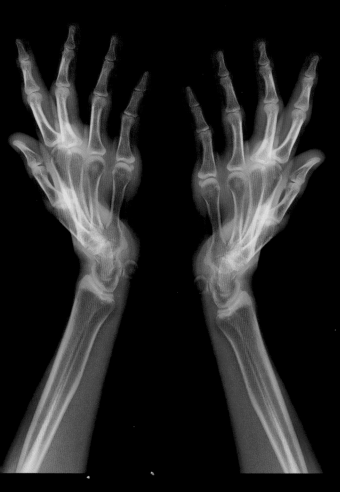

Not so long ago in some parts of West Africa, shepherds counted their flocks using shells. A shepherd would stand at his gate and as each sheep walked past, he would thread a shell onto a white strap. When 10 had gone by (one for each digit on his hands), he would remove the shells and put one shell onto a blue strap. When another 10 had gone by, he'd put another shell on the blue strap. When there were 10 shells on the blue strap, he would remove them and put one onto a red strap.

DECIMALIZATION

CHAPTER 10

In this way, he could count in hundreds, tens and units and store the results on convenient portable straps, without necessarily even knowing the names of the numbers he was using.

This ancient form of counting is known as decimal, or base 10. It uses 10 numbers (0 to 9) and counting is performed in multiples of 10 (units, tens, hundreds, thousands and so on). Like many other forms of counting in human history, it relies on tens. But this obsession with 10 may not be for any mathematical reason. Although 10s may make our sums seem easier, if we used a different number base, our sums would not really be any harder. For example, you use base 60 all the time: 60 seconds in a minute, and 60 minutes in an hour. Do you really struggle to figure out that a quarter of an hour is 15 minutes? Or that three-quarters are 45 minutes? Having 100 minutes in an hour and one hundred seconds in a minute wouldn't really change our lives much (except that minutes and seconds would be rather shorter). The reason why we use 10 for our counting is more because of the evolutionary accident that resulted in us having 10 digits on our hands. If we'd had 16 digits, we would probably think hexadecimal was the most natural way of counting. (In fact in computer science we frequently do use base 16 to count with, since it allows large numbers to be written using few digits.)

The reason why decimal is thought by many to be a simpler system of counting is because there have been some rather strange systems in use for centuries. For example, instead of counting using multiples of 1 and 10, imagine counting using multiples of 1, 12 and 20. Or even worse, what about counting using multiples of 1, 16, 14, 8 and 20? Yet this is exactly how millions counted their money for centuries in the United Kingdom, with pounds, shilling and pence (12 pence to 1 shilling and 20 shilling to 1 pound). And the second system is still very familiar to many, for it is how weights were measured, with ounce, pound, stone, hundred-weight and ton (16 ounces in 1 pound, 14 pounds in 1 stone, 8 stone in 1 hundredweight, 20 hundredweight in 1 ton). If that wasn't confusing enough, what about the measurements that count using multiples of 1, 12, 3, 220, 8 and 3? Maybe you've guessed that this is inches, feet, yards, furlongs, miles and leagues (and that's missing out some of the more obscure measures).

Weird Counting

The history of our extraordinarily diverse and often rather silly counting systems is ancient. One of the oldest we know of is that of the Sumerians, who migrated to Lower

Mesopotamia from regions such as Iran over 6,000 years ago. These peoples counted in base 60, so instead of counting: 1, 10, 100 and so on, they counted: 1, 60 and 360. There are many theories about why they would use such a bizarre method, but they lived far too long ago for us to know for sure. One idea is that they merged two earlier counting systems: one which used the digits of a hand to count using fives: 1, 5 and 25 and one that used the knuckles of one hand excluding the thumb to count using twelves: 1, 12 and 144. If the migration of people into new regions caused different cultures to blend, then it could be that the systems of counting were also merged into one. So instead of counting in fives or twelves, they began counting in five lots of 12s, or 60s.

Below: Ancient Mesopotamian clay tablet showing tally of sheep and goats using cuneiform characters.

Whatever its origins, this strange method of counting was inherited by the Babylonians, then the Greeks for their scientific numbering system, and then the Arabs, and then us. It's the reason why we still count using base 60 for our time (60 seconds in a minute, 60 minutes in an hour) and for measuring angles (60 times 60 degrees in a circle, 60 minutes in each degree and 60 seconds in each minute). The Chinese calendar also has a 60-year cycle.

Counting using base 12 (duodecimal or dozenal) is also an ancient practice that continues to this day. It's not difficult to understand why counting in 12s has always been popular (for example, the Romans used it for their fractions). Twelve has more factors than 10, so it's much easier to divide into halves, thirds, quarters and sixths. This was always very important in trade, and for any form of counting or measurement where fractions might be needed, and so 12 became incorporated into almost all of our methods of counting. It is for this reason we have 12 months, 12 signs of the zodiac and two times 12 hours in a day. It's also the reason why there are 12 inches in a foot, and there used to be 12 pence in a shilling.

The other strange numbers that we see in older counting systems are there purely for historical reasons. For example, the mile was a unit of distance first used by the Romans, and was 1,000 double steps (*mille passuum* in Latin), or 5,000 Roman feet. The Romans also had a measure called a *stade* (from which we have the word stadium), which was one-eighth of a mile, a measure taken from the Greeks. By the ninth century, the furlong was a more popular word

*Above: Roman milestone
at Capernaum Galilee.*

became a standard unit of length and was decreed to be 16.5 feet or 5.5 yards long. A furlong then became 40 rods long, and the mile became 8 furlongs (just as there had been 8 stades in a Roman mile). So it is because of the rod that the number of feet in our mile became 5,280 (8 x 40 x 16.5) instead of the easier 5,000 feet used by the Romans. Similar rather silly stories explain the strange numbers in our weights and temperature scales.

By the time Belgian mathematician Simon Stevin was born in 1548, the Imperial measurements were well established, if very confusing. As we saw in Chapter ϕ, Stevin was responsible for introducing decimal numbers to Europe. His book *De Thiende* (or, in English, *The Tenth*), in which he described how decimal fractions could be written, also stated Stevin's belief that the introduction of decimal coinage, measures and weights would only be a matter of time. He could clearly see that measuring distances or paying bills would be so much easier if everything was calculated in multiples of 10. Stevin would have been astonished to find out just how long metrification actually took, and amazed that some countries (including the United States) do not use metric weights or distances in the 21st century.

One of the key problems to be solved when producing a new metric system was not which numbers to use, it was determining how long or how heavy each unit should be. One of the pioneers, who is generally credited with the invention of the first metric system, was a French theologian called Gabriel Mouton. Born in 1618, he spent his life at Lyon, France, in the

instead of stade (the word comes from the Old English *furh*, meaning furrow, and *lang*, meaning long, and derived from the length of a furrow in 1 acre of plowed land). But there were now many measures: the inch, foot, yard, rod, furlong and mile.

Things were so confusing that around A.D. 1300, England by decree standardized many of its measures. Since Roman times there had been 12 inches to a foot and 3 feet to a yard. The rod (also known as a perch or pole) was the length of an ox goad — a long pointy stick used by medieval plowmen to make their cattle move. It

Above: Trumbull's painting
Declaration of Independence.
A vote by congress to convert
to a metric system was
defeated, so the United States
still uses the Imperial system.

church, studying mathematics and astronomy
in his spare time. Mouton decided that a metric
system for measuring length would be far
superior, and so invented his own. He suggested
that the largest unit, a *mille* (mile) should be
equal to the length of the arc of one degree of
longitude on the Earth's surface. But since he
couldn't measure such a distance, he decided
to use the properties of a pendulum instead.
Mouton knew from the work of Galileo that the
time taken for a pendulum to swing depended
only on its length. He performed experiments

and discovered that a pendulum of length one
virgula (equivalent to about 1 foot) would swing
3959.2 times in 30 minutes. This enabled him to
standardize the length of virgula, and make all
his other units of measurement multiples in
relation to this. He suggested seven new units:
centuria, *decuria*, *virga*, *virgula*, *decima*,
centesima and *millesima*, each equal to 10 of
the previous unit.

Sadly, Mouton's measurements did not
catch on. It was over a hundred years later
before France adopted the metric system, in

1795. But ironically, the United States nearly beat France to it. Thomas Jefferson had been given an English translation of Stevin's book *The Tenth*, called *Disme, The Arts of Tenths or Decimal Arithmetike*. He was so impressed by what he read that in 1783 he helped to set up their decimal currency and in 1790 he proposed a decimal measurement system for the United States. But when the U.S. congress voted on the new system of measurement, the motion was lost by a single vote. Despite Jefferson's enthusiastic support for a metric system of measurements for the rest of his life, the United States remains one of the few countries in the world to have kept the old Imperial system.

France led the development of the new decimal system and initially tried to decimalize everything, including time. Lalande was the man most dedicated to this ideal.

Jérôme Le Français was born in 1732 in Bourg-en-Bresse, France. Jérôme studied in Jesuit College in Lyon and very nearly joined the Jesuit Order. But his parents persuaded him to go to Paris and study law, which he did. By the age of 20, he had changed his name to Jérôme Le Français de la Lande, but when the French Revolution began, in order to seem less like a member of the aristocracy, he changed it to Lalande. While he studied law, Lalande also took some classes in astronomy in his spare time.

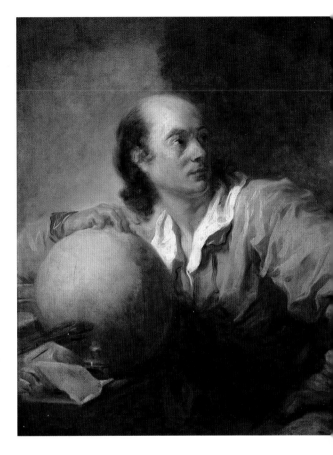

Above: Fragonard's oil painting of French astronomer Jérôme Le Français de Lalande.

When he graduated, he was offered a job with the Académie des Sciences where he helped take observations of the Moon and Mars to help determine their distances. As a result of this work, he was admitted to the Prussian Academy where he interacted with great mathematicians such as Euler. By the age of 21, Lalande had returned to France and been elected to the prestigious Académie des Sciences in Paris. Lalande's career blossomed as he made successful new predictions about the arrival of Halley's Comet. He wrote many scientific and popular books, becoming very well known. His distinctive appearance was also well known; according to one writer:

He was an extremely ugly man, and proud of it. His aubergine-shaped skull and shock of straggly hair trailing behind him like a comet's tail made him the favourite of portraitists and caricaturists. He claimed to stand five feet tall, but precise as he was at calculating the heights of stars he seems to have exaggerated his own altitude on earth. He loved women, especially brilliant women, and promoted them in word and deed.

Lalande was also an outspoken atheist, a fact that he claimed may have helped save him during the revolution. He maintained a dictionary of atheists, listing many important figures as supporters. He wrote:

It is up to the scholars to spread the light of science, so that one day they may curb those monstrous rulers who bloody the earth; that is to say, the warmongers. As religion has produced so many of them, we may hope to see an end to that as well.

But some of his contemporaries laughed at his views, suggesting that they were bitterness against God for making him so ugly. Their descriptions of him were not kind:

... his knock knees and rickety legs, his hunched back and little monkey's head, his pale wizened features and narrow creased forehead, and under those red eyebrows, his empty glassy eyes.

Luckily, Lalande seemed to take these comments in his stride, writing:

I am an oilskin for insults and a sponge for praise.

Lalande survived the Revolution and by 1791 was elected Head of the Collège de France. One of his first acts was to allow women to be taught for the first time. This was a time of great change in France; Lalande took advantage of his fame and the turbulent times to help introduce new, previously unheard of innovations.

At the Third Stroke, the Time Will Be 86 Past 5 Precisely

The Revolution was so momentous that it was decided that a new calendar was needed, to break the links to the traditional Christian-based system that had been used by the now defunct aristocracy. There were various proposals for the date of the new year I; Lalande's proposal was successful. He suggested that the first day of the first year should begin on September 22, 1792, which was the day on which the French republic was founded and coincidentally also the autumn equinox. The year

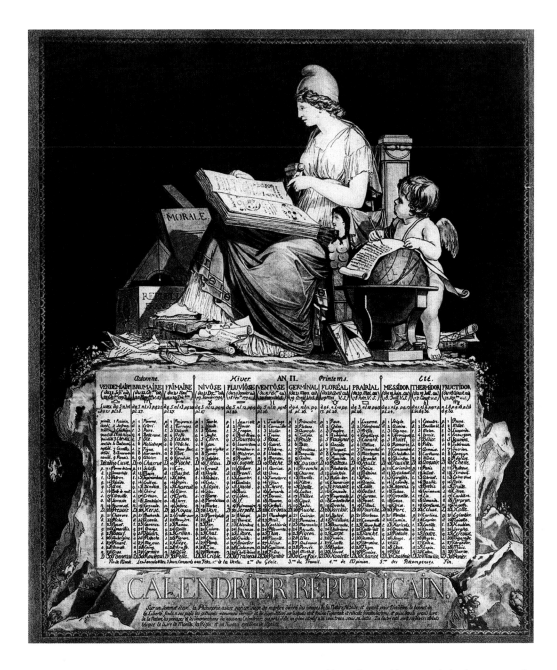

Above: The French Republican calendar issued after the Revolution of 1789. The calendar, which began on September 22, 1792, was based on the decimal system.

would now have 360 days (the extra five were holidays). There would still be 12 months in a year, but each month would be only three weeks long, each week being 10 days. To make this a little more acceptable to the working man, Lalande suggested that the middle of the week should have a holiday (a bit like a second weekend). Believe it or not, this proposal was accepted, and at the beginning of year II (autumn 1793), the new calendar became the official system in France.

This was only the beginning. By November 1, 1795 (using our calendar), a new law had been passed, dictating that time and angles should now

Above: Portrait of French mathematician Pierre-Simon Laplace.

making a right angle 100 gradians. So the Earth would rotate 40 gradians each hour instead of 360 / 24 = 15 degrees of the old system. It wasn't going faster; the hours were longer and gradians smaller than degrees. (If you have a scientific calculator, you may find that it has a grad option — this allows you to calculate in gradians instead of degrees.)

It was suggested that the wigmakers could help with the enormous task of calculating new calendars, timetables and trigonometry tables. There were rather a lot of unemployed wigmakers, for the main wearers of wigs — the aristocracy — had lost their heads during the Revolution (and those who avoided the trip to the guillotine didn't want to advertise the fact by wearing a wig).

The French mathematician Laplace relished the new system of time, and had a new watch made with 10 hours on its dial. He wrote a five-volume book of mathematics which used the new units of time and angle. But Laplace was unusual in his acceptance of the system. Most people in France found the whole idea difficult to accept and horribly confusing to adjust to. Ten years of 10-hour days and 100-minute hours passed before Napoleon Bonaparte scrapped the system to ensure the Church would support him. In the end, Laplace assisted Napoleon, suggesting that there were scientific flaws in the new calendar.

But although the clocks and calendars were only briefly metric, distances and weights were another matter. By 1795, a new measure of distance had also been introduced to France. From the Greek word *metron*, meaning a measure, the unit of distance was called the

be decimal. There were now 10 hours in each day, 100 minutes in each hour and 100 seconds in each minute. The new law required new clocks to be made and used immediately. Angles were also different. Now there would be 400 degrees (or gradians, as they became known) in a circle,

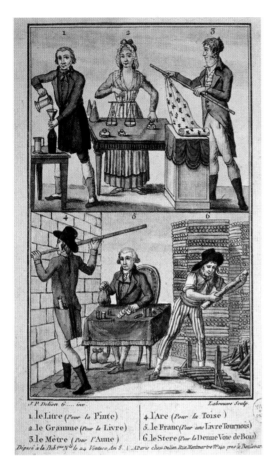

1. le Litre (*Pour la* Pinte)
2. le Gramme (*Pour la* Livre)
3. le Mètre (*Pour l'*Aune)
4. l'Are (*Pour la* Toise)
5. le Franc(*Pour une* Livre Tournois)
6. le Stere(*Pour la* Denue Voie de Bois)

Above: Illustration showing the institution of the metric system in France.

is defined in relation to the speed of light in a vacuum and so is not subject to change (1 meter is the distance traveled by light in a vacuum in 1/299,792,458 of a second). The meter, and all of its cousins (the centimeter, the millimeter, the kilometer) were an unstoppable idea.

Once the meter was established, volume and mass followed. One liter was defined to be the volume of a cube with sides of 10 cm (leading to milliliters, deciliters and so on). One gram was defined to be one cubic centimeter of water at its maximum density (4°C). Later

Above: Illustration showing the 10-hour pocket watch of the metric system in France.

meter. It was initially suggested to be the length of a pendulum that had a half-period of one second, and then became one ten-millionth of the distance from the North Pole to the equator, along the meridian running near Dunkirk in France. This distance was calculated (slightly incorrectly as it turned out), and a brass bar was made to the exact length. Later a platinum bar was made — it was slightly more stable at different temperatures (and 100 years later a platinum-iridium bar was made). Today the meter this also became standardized in a block of platinum-iridium. With the gram defined, we also had the kilogram and the metric ton. One interesting difference between pounds and kilograms is that one is a measure of weight, the other is a measure of mass. On Earth, this doesn't make much difference to us, but if we were to make the same measurements on the Moon then the difference is very important. The Moon has less gravity, so the same mass has less weight. A 196 pound person would weigh less than 42 pounds on the Moon. An 80 kilogram person would still have a mass of 80 kilograms on the Moon. (Your bathroom scales would get this wrong, for they estimate your mass based on your weight.) We normally don't worry about this difference very much, but when sending rockets into space, the difference between mass and weight becomes very important indeed.

The metric system of measures soon spread around the world, and most countries have now adopted it. Even those who still use the Imperial system are actually using a system that is based on the metric system, for in 1958 it was internationally agreed that all measurements (the inch, the pound, etc.) be an agreed proportion of the metric measures. So according to the official Système International d'Unités, the yard is now officially 0.9144 meters, and the pound is 0.45359237 kilograms. These are exact amounts, so although we may use the same names, our imperial measures are not the same as those used in previous centuries. Today an inch is the name for 2.54 centimeters and nothing else, except perhaps the worm.

Right: The Pythagoreans believed in the power of the tetrad, or number four. This painting represents the four seasons.

Sacred Tetractys and Triangles

Ten may have changed our world through decimalization, but 2,000 years ago it was the mystical properties of the number itself that made 10 central to the religion of the Pythagoreans. According to one ancient writer:

> Ten is the very nature of number. All Greeks and all barbarians alike count up to ten, and having reached ten revert again to the unity. And again, Pythagoras maintains, the power of the number 10 lies in the number 4, the tetrad. This is the reason: if one starts at the unit (1) and adds the successive number up to 4, one will make up

the number 10 (1 + 2 + 3 + 4 = 10). And if one exceeds the tetrad, one will exceed 10 too ... So that the number by the unit resides in the number 10, but potentially in the number 4. And so the Pythagoreans used to invoke the Tetrad as their most binding oath: *By him that gave to our generation the Tetractys, which contains the fount and root of eternal nature ...*

If numbers were the key to understanding the universe, as the Pythagoreans believed, then 10 was surely special. The magic relationship between the first four numbers and 10 led them to create a whole philosophy based on 10 sets of 4. The Sacred Tetractys was their view of how the world could be explained and understood:

#		
1	*Numbers*	1+2+3+4
2	*Magnitudes*	point, line, surface, solid
3	*Elements*	fire, air, water, earth
4	*Figures*	pyramid, octahedron, icosahedron, cube
5	*Living things*	seed, growth in length, in breadth, in thickness
6	*Societies*	man, village, city, nation
7	*Faculties*	reason, knowledge, opinion, sensation
8	*Seasons*	spring, summer, autumn, winter
9	*Ages of a person*	infancy, youth, adulthood, old age
10	*Parts of living things*	body, three parts of the soul

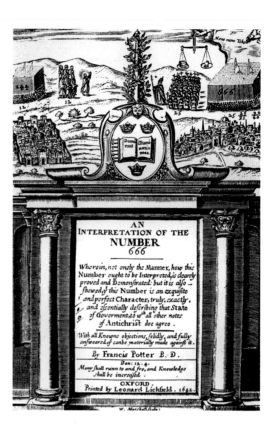

This was the list of their 10 commandments that guided their lives, except these were also explanations that helped inquiring minds to explore and understand the truth, as opposed to constrain and restrict unwanted behavior. Numbers formed the heart of this philosophy so it was no coincidence that the Pythagoreans revered 10 sets of 4. When the first four numbers are written one below the other using dots instead of Arabic symbols, they form a perfect triangle, and the total number of dots comes to exactly 10:

Above: Illustration of an interpretation of 666, the "number of the beast," from 1642.

This is why 10 is said to be the fourth triangular number: it makes a triangle out of dots. If you try writing numbers using rows of dots, one below another, you'll quickly discover that the first 10 triangular numbers are 1, 3, 6, 10, 15, 21, 28, 36, 45, 55. All form perfect triangles of dots, like bricks stacked on top of each other.

Triangular numbers are very easy to calculate: you simply add the natural numbers together. So the first triangular number is 1, the second is 1 + 2 = 3, the third is 1 + 2 + 3 = 6, and the

fourth is 1 + 2 + 3 + 4 = 10. It turns out that all perfect numbers are triangular (if you recall from Chapter 1, a perfect number is one of those very rare numbers that can be formed by adding up all the smaller numbers that make up its divisors). One of the most famous triangular numbers is the so-called "number of the beast": 666. (There is now some question about 666 being the number of the beast — it turns out that it may have been a mistake made centuries ago in copying of the scriptures. In the oldest surviving copy of the New Testament — some 1,500 years old — the true "evil" number appears as 616. Oops!)

But while triangular numbers have been known for thousands of years, it was not until the 17th century before one of the most in-depth investigations was made.

Blaise Pascal was born in 1623, at Clermont, France. His mother died when he was only 3 so he was brought up and educated by his father (a lawyer and amateur mathematician), who had some rather unorthodox ideas about teaching. His father believed that Blaise should not be educated in mathematics before the age of 15, and so all mathematics books were removed from the house. Perhaps this was a clever ploy by his father or perhaps it was accidental, but Blaise soon became curious about this banned subject and taught himself geometry in his spare time. By the age of 12 he had discovered that the sum of angles of a triangle equaled two right angles (180 degrees). When his father discovered his son's work, he relented and gave Blaise the work of Euclid to read. Before long Blaise was accompanying his father to meetings of mathematicians, where he presented his own theorems on geometry.

When he was 16, his father was given a job as a tax collector, so the family moved to Rouen. A year later Blaise Pascal had published his first work on geometry. By the age of 22 he had invented a mechanical calculator (which he called the *Pascaline*) to assist his father in the calculation of currency. This was made rather difficult by the French currency, which like the British was based on 12 and 20 (12 deniers in a sol, 20 sols in a livre). A year later his father broke a leg and was nursed by two brothers from a nearby religious order. Pascal was profoundly affected and became deeply religious, choosing to "contemplate the greatness and the misery of man."

Below: Illustration of Blaise Pascal.

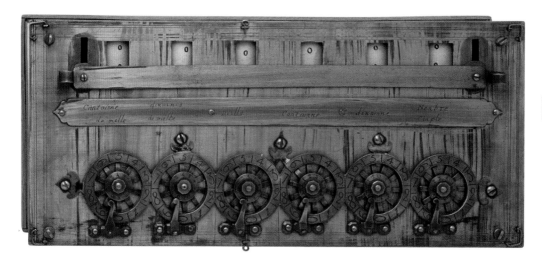

Above: Calculator created by Blaise Pascal in 1642. The holes at the top of the calculator would show solutions to calculated numbers.

Nevertheless, he also found time to continue his research, and next looked at atmospheric pressure. Before long, Pascal had demonstrated to his satisfaction that a vacuum could exist (a volume with zero atmospheric pressure within it). Descartes came to visit, but did not believe Pascal, later writing to a friend that Pascal:

has too much vacuum in his head.

Pascal was also described as "a man of slight build with a loud voice and somewhat overbearing manner," and "precocious, stubbornly persevering, a perfectionist, pugnacious to the point of bullying ruthlessness yet seeking to be meek and humble."

Luckily insults did not slow Pascal, who showed a few months later that atmospheric pressure decreased at higher altitudes. This led him to deduce that there must be a region above the atmosphere which was a vacuum (or space, as we now call it). His later work explained pressure in liquids, geometry, probability and explored various philosophical and religious ideas. He even attempted to combine everything to prove that a belief in God was rational. He used probabilistic and other mathematical arguments to show that:

If God does not exist, one will lose nothing by believing in him, while if he does exist, one will lose everything by not believing.

This became known as Pascal's wager — his conclusion was that "we are compelled to gamble." But his arguments were flawed. (For example, we cannot choose our beliefs as we choose fruits from a tree and if we could we would lose our integrity; nor can we ever know the consequences of a lack of belief in God.)

Pascal's triangle is a very remarkable arrangement of numbers. It has a "skin" of 1s going down both diagonals. The next diagonals in are the natural numbers in order. Next to those are two diagonals comprising the triangular numbers in order. Moving inward, next to those are diagonals containing the pyramidal triangular numbers (or tetrahedral numbers) in order. (To make pyramidal triangular numbers, just stack your dots into a 3-D pyramid instead of the 2-D triangle we used for the triangular numbers.) Moving inward again, the next diagonals are the pentatope numbers in order, and so it goes on … It's also possible to find prime numbers, Fibonacci numbers, Catalan numbers, and if you color all odd and even numbers black and white, you get a fractal

Although Pascal's philosophical views were debatable, his work in mathematics was groundbreaking. One area that he helped to explore better than any before him was number theory and triangular numbers. Because of this, though he was not its inventor, one particular pattern of numbers became known as Pascal's triangle (see box, below).

Pascal's triangle

You can make Pascal's triangle very easily. Begin with a 1 as the peak of the triangle. Now build the "bricks" under it, by following the simple rule: the current number is the sum of the two above it (one top-left and one top-right). If there's only one number above, then treat the missing number as a 0. The result is a very special triangle of numbers:

```
                              1
                          1       1
                       1      2       1
                    1      3       3      1
                 1      4       6       4      1
              1      5      10      10      5      1
           1      6      15      20      15      6      1
        1      7      21      35      35      21      7      1
     1      8      28      56      70      56      28      8      1
  1      9      36      84     126     126     84      36      9      1
1     10     45     120     210     252     210     120     45     10     1
1    11    55    165    330    462    462    330    165    55    11    1
1    12    66    220    495    792    924    792    495    220    66    12    1
1    13    78    286    715   1287   1716   1716   1287   715    286    78    13    1
1    14    91    364   1001   2002   3003   3432   3003   2002   1001   364    91    14    1
```

shape known as a Sierpinski triangle (we'll see more of fractals in Chapter i). Pascal's triangle also enables us to expand special equations known as binomials.

Pascal became more religious as he aged. On one occasion, the horses pulling his carriage bolted, and he was left hanging on the edge of a bridge over the Seine. He was rescued unharmed, and attributed his survival to God and wrote a religious verse that he carried for the rest of his life in his jacket. Later in his life, while working on another mathematical problem, he suffered from insomnia and toothache when the solution occurred to him — and suddenly his pain was gone. He attributed this to a divine intervention to proceed with the idea and spent the next eight days working continuously on it. Pascal's writings on religion became very influential and were seen as setting new standards in French prose, using

Binomials

A binomial is simply a little equation that has two elements in it, for example:

$(x + 1)^2$

It often helps to expand binomials in order to figure out (or even just approximate) the value of the expression. Expanding the binomial above gives:

$x^2 + 2x + 1^2$

which, if you're observant, you'll have noticed appears in the third row of Pascal's triangle:

1   2   1

Rather amazingly, it turns out that this works for any binomial. So to expand the general binomial:

$(x + y)^n$

you need to find the coefficients a_0, a_1, a_2, ..., a_n in the expanded expression:

$a_0 x^n + a_1 x^{n-1} y + a_2 x^{n-2} y^2 + ... + a_{n-1} x y^{n-1} + a_n y^n$

and the coefficients will be precisely the numbers on row n+1 of Pascal's triangle. Just to prove it works, let's try this one:

$(x+4)^4 = a_0 x^4 + a_1 x^3 \times 4 + a_2 x^2 \times 4^2 + a_3 x^1 \times 4^3 + a_4 \times 4^4$

On row 4+1=5 of Pascal's triangle, we have the numbers: 1, 4, 6, 4, 1, so the expansion is:

$1x^4 + 4x^3 \times 4 + x^2 \times 4^2 + 4x^1 \times 4^3 + 1 \times 4^4$

or, to make things look much less complicated:

$x^4 + 16x^3 + 96x^2 + 256x + 256$

humor and cutting criticism to make the points eloquently. He wrote one book as a series of fictional letters from a Parisian to a friend in the provinces. *Letter XVI* contained the memorable apology: "I would have written a shorter letter, but I did not have the time."

In the last few years of his life Pascal was in great pain as a tumor grew in his stomach. He gave up science and spent his time giving to the poor. Pascal died at the age of only 39 from a brain hemorrhage. He was not forgotten by the world, however. As well as Pascal's triangle, in 1968 the PASCAL computer programming language was named after him.

Right: Pascal's triangle as it was first printed in 1527 as a title page to the arithmetic of Petrus Apianus.

Numbers, numbers everywhere. Yet why are we so unoriginal at picking them as our lucky or unlucky numbers? Why do we tend to choose the same numbers, even when we think we're choosing randomly? Perhaps there is something about the numbers themselves that is affecting us. Can there really be such things as lucky or unlucky numbers? Can superstition have some truth behind it?

TRISKAIDEKAPHOBIA

CHAPTER 12a

Think of a number between 1 and 100. Got one? Now see if I guessed it correctly[1]. In 2006, the website www.arandomnumber.com was created by Greg Laabs, from California, that asked the same question, with no clues of what it was for. At the time of writing he had received 71,618 numbers from people. The results (which he collated especially for you, the reader of this book) were fascinating. The top five numbers (typed by people who thought of the number before typing it in) were, in order of popularity: 5, 7, 37, 56 and 42. Number 5 was 3 times more common than it should have been, if everyone had picked a number perfectly randomly.

The selected numbers suggest an interesting mix of influences on people. The number 5 is in the center of the row of numbers and the number pad on a computer keyboard, making it very visible and easiest to hit quickly. Similarly 56 is very easy to type quickly. Numbers 7 and 37 are more interesting, for they are often considered to be the most commonly chosen numbers, perhaps because they are prime or they seem somehow lucky or balanced. And the choice of 42 by a significantly higher than normal percentage of people almost certainly shows the influence of *Hitchhiker's Guide to the Galaxy* by Douglas Adams. (Read the book if you want to know why 42 is so special.) The five least popular numbers, picked by almost nobody at all were (with the least favorite last): 40, 91, 94, 70 and 90. For some reason, few people related to these numbers — perhaps no one thought they were lucky, interesting or special.

As Greg's experiment illustrates, people aren't good at picking random numbers. For some reason we choose some numbers more often than others. Whether superstition or other cultural influences, people are not very unique.

Be Careful What You Believe

What is superstition, anyway? The origin of the word superstition is thought to be from the Latin word *superstes* (from *super*, meaning over or beyond, and *sto*, meaning to stand). This word had two meanings: someone who witnesses something, and someone who survives something. So *superstitio* became a way of telling events as if you had been there and survived them. It became associated with diviners who pretended to predict the future. Today superstition has evolved into a set of rules for the prediction of the future. If you walk under a ladder, something bad will happen to you. If you see a shooting star and make a wish, it will come true. If you sit on a seat with a number 13 on it, you will have bad luck.

Thirteen is a number that typifies superstition. Many people are nervous of anything with a 13 on it. If any number could have some mystical property to do with luck, it would surely be 13.

[1] I guess 37. To find out why, keep reading the main text

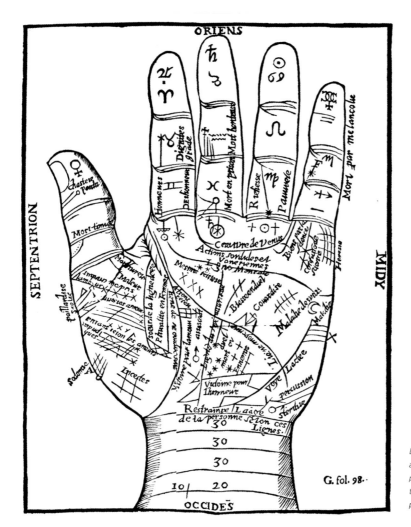

Left: Superstition became associated with diviners who pretended they could predict the future — one such common practice involved palm reading.

The origin of 13 being associated with bad luck is lost in the distant past. There are many theories, some about Viking gods, some about the number of disciples at the last supper, others about execution of Templars knights, and some even suggest that it shows an ancient prejudice against women because there are 13 lunar (menstrual) cycles in a year. We do know that Hindus considered it unlucky for 13 people to gather in one place, and it has been said that the Turks so disliked the number 13 that they almost deleted it from their vocabulary. But we also know that 13 is lucky for the Chinese, and was also thought to represent good fortune for the ancient Egyptians. There is no doubt that 13 is not a well-liked number in the West, with many cities having no 13th street or 13th avenue, buildings having no 13th floor and houses skipping the number 13. Even in the age of computers and jet engines, the seating layouts of many passenger aircraft have no row 13. (There are some people who may even find this paragraph frightening, for it contains the number 13 repeated 13 times.)

Those with a fear of the number 13 are called triskaidekaphobes. These people will point out that some quite notorious criminals had 13 letters in their names (for example, Jack the Ripper, Charles Manson, Jeffrey Dahmer, Theodore Bundy and Albert De Salvo). They'll say there are 13 witches in a coven (whatever that is supposed to mean to us).

However, today superstitious belief such as this does not make any rational sense and most people will not be able to justify their superstitious fears. Like our celebrations of Halloween or even Christmas, when a practice becomes embedded in our culture it is slowly transformed from habit to ritual to tradition. So in the western world, Friday the 13th has become traditional. Unfortunately traditions can have teeth. Whatever the truth behind luck or unluckiness, enough people notice this "special" date to make it effect their behavior. Rather remarkably, a scientific article published in the *British Medical Journal* in 1993 showed that hospital admissions due to car accidents increased by as much as 52 percent on Friday the 13th compared to one week earlier. This was despite the fact that noticeably fewer people chose to drive on the 13th compared to the 6th. Not only that, but in 2005, ABC News reported that businesses in America lose nearly 1 billion dollars every Friday the 13th because of superstitious workers who stay at home, or people cancelling their travel plans on airplanes or buses. The fear of this date is so common that there are not one, but two words to describe people with such a phobia: *paraskevidekatriaphobe* and *friggatriskaidekaphobe*. Thankfully they're both sufficiently difficult to pronounce that

Above: Superstitious initiation: Members are forced to walk under a ladder, which is said to bring bad luck — especially given the presence of the number 13.

they're unlikely to be used as the title of a movie (as was arachnophobia, or the fear of spiders).

Ironically, for many people a belief in unlucky Friday the 13th is enough to make the day unlucky for them. Enough people become nervous and overreact on this date that it really can be a more dangerous time to travel by car. It's not the number that makes the day unlucky, it's how we react to it. In Australia, where they "live upside down," there's another interesting trend: on Friday the 13th far more lottery tickets are bought than on other days. Australians clearly think that they may be more lucky on this date.

Above: Breaking mirrors is supposed to be unlucky, as is indicated by the number 13, and is thought to bring seven years of bad luck.

Mathematics of Luck

Numbers and luck are related, but not in the way that superstition suggests. Luck is basically good fortune, so how do you know if you will be lucky? What are the chances of you winning the big prize, or getting that job you wanted?

The first two mathematicians to consider the mathematics of chance lived in the 17th century. They were Blaise Pascal (who is known for the triangle of numbers named after him, as we saw in Chapter 10) and Pierre de Fermat (known for his missing last theorem and hated by Descartes as we saw in Chapter $\sqrt{2}$). Pascal was asked to consider a gambling problem by French nobleman Chevalier de Méré. The problem was about throwing a pair of dice 24 times — would it be profitable to bet that at least one double-six appears? What are the chances of getting a double-six in 24 throws?

Pascal was friends with Fermat, and so wrote to him with the problem in 1654. The two mathematicians quickly began an excited exchange of letters, in which they invented what we now call probability. (All but the first letter still exist, and so they provide a fascinating insight into the minds of two mathematicians as they correct each other and come to agreement on new ideas.)

So how can you calculate how likely something will be in the future? How can you figure out if you'll be lucky? Imagine a slightly simpler problem. Pascal and Fermat are sitting in a Parisian café, playing a simple game of "flip a coin" together. If the coin lands heads up, Fermat wins a point. If the coin lands tails up,

Pascal wins the point. The overall winner is the first one to get 10 points. Both Pascal and Fermat put down 50 francs, so the winner will get the whole 100 francs. They play this game for a while, until Fermat is winning 8 points to 7. However, Fermat suddenly receives an urgent message that a friend is very ill and he must immediately leave. Pascal understands, but after Fermat has left he realizes he still has the whole 100 francs. He writes to Fermat and asks how the money should be divided between them. Fermat replies, showing in his letter how the game could have finished, and explaining how the money could be divided fairly.

Below: Pascal and Fermat played their game of luck by gambling on the flip of a coin.

Pascal's letter to Fermat

Dearest Blaise,

As to the problem of how to divide the 100 francs, I think I have found a solution that you will find to be fair. Seeing as I needed only 2 points to win the game, and you needed 3, I think we can establish that after four more tosses of the coin, the game would have been over. For, in those four tosses, if you did not get the necessary 3 points for your victory, this would imply that I had in fact gained the necessary 2 points for my victory. In a similar manner, if I had not achieved the necessary 2 points for my victory, this would imply that you had in fact achieved at least 3 points and had therefore won the game. Thus, I believe the following list of possible endings to the game is exhaustive. I have denoted "heads" by an "h," and tails by a "t." I have starred the outcomes that indicate a win for myself.

h h h h *	h h h t *	h h t h *	h h t t *
h t h h *	h t h t *	h t t h *	h t t t
t h h h *	t h h t *	t h t h *	t h t t
t t h h *	t t h t	t t t h	t t t t

I think you will agree that all of these outcomes are equally likely. Thus I believe that we should divide the stakes by the ration 11:5 in my favor, that is, I should receive $(^{11}/_{16})*100 = 68.75$ francs, while you should receive 31.25 francs.

I hope all is well in Paris,
Your friend and colleague,
Pierre

(The original letters between Pascal and Fermat were often a little more complicated than this, but you get the idea.) Between them, these two mathematicians realized that fractions and ratios could be used to inform us how likely something was. So the chances of a coin landing heads up is 1 in 2, or a probability of $\frac{1}{2}$, because a coin has only two sides and it is equally likely to land on either side. They also realized that addition and multiplication could be used with probabilities. So the probability of throwing a 6 on a die twice is $\frac{1}{6} \times \frac{1}{6}$ ("and" works like multiply). The probability of throwing a 6 or a 3 on a die is $\frac{1}{6} + \frac{1}{6}$ ("or" works like addition). Using these and many other similar relationships, we can work out the probability of many events — a fact exploited by all the casinos, betting shops and racing events around the world to ensure that we lose a little more often than we win, and therefore make their businesses enormously profitable. We can also work out the probability of the original problem posed to Pascal (see box opposite).

The mathematics of probability are very useful, but they will not always work. Probabilities tell us the most likely outcomes given a set of assumptions (that the dice are fair, or the horse is fit, or the coin actually has both a head and a tail on it). All too often the assumptions are just a little bit wrong, so in the real world you can only use them as a guide. Sometimes there is a possibility that you can beat the odds ... if you're very, very lucky.

Below: Casinos exploit the rules of probability by ensuring that gamblers lose more often than they win.

Finding Meaning with Numbers

Superstitions (or numerology) might deceive you into believing that a number might be significant, or that some meaning of the number 13 causes a change of luck. Probability allows numbers to signify how lucky you may be. Other meaningful numbers are not hard to come by. Through

Probability of problem posed to Pascal

We want to know the probability of at least one double-six in 24 throws of two fair dice. (Fair dice are dice that are not weighted in a way that makes them unfair.)

We know that the probability of throwing one 6 is $\frac{1}{6}$ (there are six sides on a die, and only one is a 6). So the probability of throwing one 6 AND another 6 is $\frac{1}{6}$ x $\frac{1}{6}$ = $\frac{1}{36}$

The probability of *not* throwing a double-six must be all the other outcomes, or $1 - \frac{1}{36} = \frac{35}{36}$

So the probability of *not* throwing a double-six 24 times in a row is $\frac{35}{36}$ x $\frac{35}{36}$ x $\frac{35}{36}$ x ... x $\frac{35}{36}$ = 0.508596

The probability of throwing at least one double-six (which means one or more double-sixes) must be all the other outcomes, or $1 - 0.508596 = 0.4914$

We learn from this that it is slightly less likely for a double-six to appear (0.4914) than it is for it to appear (0.508596) after 24 throws (but it's very close). So if you were betting, you should bet that the double-six will not appear and you will be slightly more likely to win. Your other alternative is to make sure the two dice are thrown more than 24 times, because the chances of getting a double-six become more likely for 25 or more throws in a row. Once you're throwing 50 times, the chances of getting a double-six become 0.7555 (or 76 percent likely). And if you can convince your opponent to throw the two dice 100 times, the probability of getting a double-six is very high at 0.94 (or 94 percent likely).

cultural influences, numbers have come to have new meanings, for example 1984 (from George Orwell's famous novel *Nineteen Eighty-Four*), and Catch-22 (from Joseph Heller's novel with the same name). There are even expressions that use the word "number" to mean something new: "your number's up" (from the Bible, Daniel 5, when the King of Babylon heard that the time of his kingdom was over).

However, there are some who believe that numbers can help us to discover meaning in surprising places. In 1984, three Israelis, Doron Witztum, Eliyahu Rips and Yoav Rosenberg claimed to have found something extraordinary in the Torah (the Hebrew text of Genesis). They claimed that biographical information about medieval rabbis is "encoded" in the Torah using patterns of letters. They suggested that the names of the rabbis could be found close to the correct dates of birth and death, hidden within the text. The information could be decoded by following an Equidistant Letter Sequence (ELS), where the key letters were a certain number of characters apart. If you can find the number, then you can decode the message. For example,

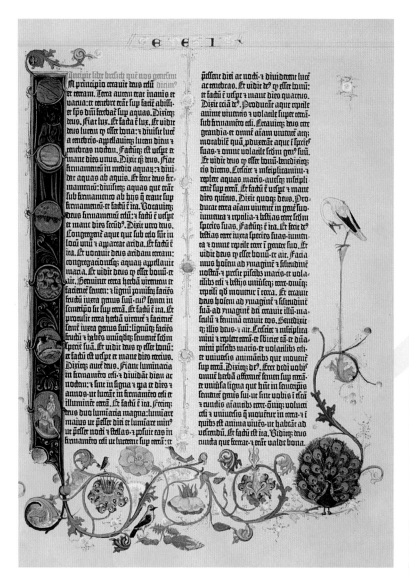

Left: Page taken from Genesis, from the Gutenberg Bible. Numerology has often been applied to the Bible.

the bold letters in **t**his **s**ent**e**nce **f**orm **a**n ELS. The hidden word is SAFEST (the skip number is –4; spaces and punctuation are ignored).

The claims were published in a journal of statistics 10 years later and caused a storm of controversy. Those who believe that hidden codes do exist in the Bible or the Torah (such as Michael Drosin in his book *The Bible Codes*) have gone on to find extraordinary predictions of disasters, the end of the world and names of important figures who had been assassinated in history, all "locked up" in the text.

As nice as the idea may seem, when mathematicians examined the claims, things didn't look so convincing. It turns out that with the right numbers and enough text you can find just about anything you want to find. Like seeing faces in clouds, patterns that may seem meaningful can be found in almost anything, but this does not make them mean anything.

In an attempt to justify his book (or perhaps just to help its sales) Drosin said, "When my critics find a message about the assassination of a prime minister encrypted in *Moby Dick*, I'll believe them."

*Above: Ancient Hebrew script.
These scripts are thought by
some to have hidden messages
encoded within them using
patterns of letters.*

A key critic and mathematician called McKay
promptly produced an ELS analysis of *Moby Dick*
predicting not only Indira Ghandi's assassination, but
the assassinations of Martin Luther King Jr., John F.
Kennedy, Abraham Lincoln and Yitzhak Rabin, as
well as the death of Diana, Princess of Wales.
McKay also worked with Australian television
personality John Safran and showed that evidence
of the 9/11 terrorist attacks could be found coded in
the lyrics of Vanilla Ice. Another mathematician
called David Thomas did an ELS on Genesis and
found the words "code" and "bogus" near to

each other 60 times. Thomas also did an ELS
analysis on Drosnin's second book, *Bible Code II:
The Countdown* (2002) and found the message
"The Bible Code is a silly, dumb, fake, false, evil,
nasty, dismal fraud and snake-oil hoax." With
that he convincingly proved that if you look hard
enough, you can find whatever message you
want to.

Numbers can be used to find messages in
text — just don't expect the messages to mean
anything. We don't need superstitious and false
new meanings; numbers are fascinating in their
own right, whether prime, perfect, real, complex
or imaginary. Numbers already have profound
and true meanings that enable us to explore and
explain our universe. It's probably best to stay
away from the scaremongerers and charlatans
who make their money from misinformation.

Numbers act like pillars, holding up our universe and making it the right shape. If π wasn't 3.14159… then all the circles and curves in the universe would be a different shape. If φ wasn't 1.61803… then all the geometric shapes, ratios and curves would be different.
If e wasn't 2.71828… then the relationships between position, speed and acceleration would be very different. These numbers are built into our universe as inextricably as space and time. They're not the only ones. There's one number that is so

AS FAST AS YOU

CHAPTER c

important that it transformed our whole view of existence. That number is c, also known as the speed of light in a vacuum.

Why should a speed be so important? It was once thought that the speed of sound was an unbreakable limit, but we soon learned better. The speed of sound through air is 742.3 miles per hour (or 331.4 meters per second), but this depends on the temperature of the air (it's faster in warmer air). We now know that it's possible to travel faster than the speed of sound. When a jet plane moves that fast, all that happens is we hear a sonic boom (caused by a shockwave of the plane moving so quickly). If you're watching from the ground, the engine noise always seems to come from quite a distance behind the place where you see the plane (caused by the fact that light travels faster than sound, so the delay between something happening and hearing it is much longer than the delay between something happening and seeing it).

But if we can travel faster than the speed of sound, then light must be exactly the same. It's true that light is much, much faster. In fact it's 186,282.397 miles per second (or 299,792,458 meters per second). This is equivalent to 670,616,629.384 miles an hour. But, if we had a big enough rocket strapped to an aircraft or spaceship, we'd be able to accelerate until we were going faster than this

CAN GO

speed. Wouldn't we? Surely, if we had a rocket with the power of, say, the Sun, we would be able to accelerate up to 190,000 miles a second? If that wasn't enough, how about a rocket with the power of a million suns?

Perhaps surprisingly, we can't do it. No matter how big a rocket we use, no matter how much we try and accelerate, we can *never* go faster than the speed of light. Our universe has a speed limit that prevents anything from going faster. The speed of light in a vacuum is as fast as you can go. It took a genius named Albert Einstein to figure out why, but before he could, we had to understand the speed of light itself.

Seeing c

For thousands of years, the very concept of a speed of light was thought to be absurd. From Aristotle to Kepler to Descartes, most believed that light was simply instantaneous. Galileo (the one who made the first telescopes to be used for studying the night sky) was the first scientist to suggest an experiment that might determine just how fast light is. He and his assistant took two shuttered lamps. They began by standing close to each other, Galileo uncovering his lamp and his assistant immediately uncovering the second lamp when he saw the light from Galileo's lamp. When standing close to each other they could judge the delay caused by

Above: The Trial of Galileo
(1633). Despite innovations
working with the speed of
light, Galileo was forced to
face the Inquisition.

human reaction times. Then they both went to
stand on the top of distant hills and did the same
experiment: Galileo uncovering his lamp and
watching for the light from his assistant's lamp.
Galileo reasoned that if light behaved like sound
and had a noticeable delay, then he should be

able to see a difference in the timing (the
assistant would have to wait for the light from
Galileo's lamp to reach him, and then Galileo
would have to wait for the light of his assistant's
lamp to come all the way back). It's clear that if
the same experiment was performed with sound
— perhaps Galileo shooting a gun in the air, with
his assistant firing another gun as soon as he
heard the shot — there could be quite a long
delay of several seconds if they stood on
different hills. But perhaps predictably, light

*Above: Rømer's observations of
Jupiter's moon, Io, enabled him
to calculate the speed of light
with some degree of accuracy.*

travels much too quickly to be tracked using
Galileo's idea. When he performed his
experiment with the lamps, he couldn't see
any noticeable difference compared to when
they stood next to each other. He concluded
that light must be at least 10 times faster than
sound, but had no real idea how fast it was.

It was not for another 50 years, in 1676, that
astronomer Ole Rømer managed to calculate the
speed of light. Rømer was born in 1644 in
Aarhus, Denmark. He studied at Copenhagen
University and was taught by Rasmus Bartholin
(a scientist who studied the refraction of light).
He then took a job at the Paris observatory,
where he was employed to make observations
of the planets and their moons. It was during
this work that he noticed something strange.

Rømer was observing the motion of Io, one
of Jupiter's moons. He could see that Io orbited
Jupiter about every 42.5 hours. But he noticed a
peculiar discrepancy in his observations. When
Jupiter and Io were furthest away from the
Earth, Io seemed to take a little longer to
emerge from the shadow of Jupiter. When
Jupiter and Io were closest to the Earth, Io
seemed to emerge from Jupiter's shadow a little
earlier. Rømer realized that the distance

between the Earth and the Jupiter-Io pair
seemed to be affecting the timing of Io's orbit.
It was only a few minutes difference, but even
with nothing but simple telescopes, logarithm
tables, pen and paper to make the calculation,
Rømer was able to detect this difference.

The change in distance between Jupiter and
the Earth was not surprising. Since Kepler it had
been known that a planet farther from the Sun
would orbit slower. This meant that for every
one orbit by the Earth around the Sun, Jupiter
only moved a fraction of its orbit. Like a faster
racing car lapping a slower one on a circular
track, sometimes the Earth would pass "close
by" Jupiter, and sometimes the Earth would be
on the other side of the track (orbit) from Jupiter.

Right: Engraving of Ole Rømer using his Meridian Circle.

Rømer knew that the position of the Earth could not be affecting Io directly — it never came close enough for that to happen. The only other thing he could think of was that his observation was being affected by the distance. If light had a finite speed, then the extra distance it had to travel when the Earth and Jupiter were far apart would cause a delay. He would have to wait several minutes longer for the light to reach Earth (and his telescope). Rømer calculated that light took about 22 minutes to cross the orbit of Earth and was able to predict how late Io would appear. (In fact he got the number wrong — it actually takes about 17 minutes for light to cross this distance, so Rømer thought light traveled a little slower than it actually does.) Nevertheless, despite being slightly inaccurate, Rømer was the

62 384

first scientist to provide concrete evidence that light is not instantaneous.

Rømer returned to Copenhagen in 1681 and became a professor of astronomy. He made many important contributions. These include inventing the Meridian Circle and other instruments to help position telescopes accurately. He also worked for the King and helped introduce the first standardized system of weights and measures, and the Gregorian calendar to Denmark. He was made the second Chief of Copenhagen Police in 1705 and helped improve his city by trying to assist the poor, unemployed and prostitutes, as well as by providing a good public water supply. He even invented the first street lights in Copenhagen (oil lamps). Two years before his death in 1710, a man called Fahrenheit visited Rømer. The result was a temperature scale that is still in use today.

Despite his eventual fame, Rømer's results on the speed of light were not believed by everyone, and the following decades were filled with heated debate on the subject. A man called James Bradley was to resolve the issue in 1728.

Bradley was born in 1693 in Sherborne, Gloucestershire, England. He was influenced by his uncle the Reverend James Pound, who was an astronomer, and he assisted Pound by making observations in the rectory of Wanstead in Essex. By the age of 25 Bradley had published details of his own observations and was elected

a fellow of the Royal Society soon after. But instead of a career in astronomy, he decided to enter the church and was ordained in 1719, becoming the vicar of Bridstow in Monmouthshire. In his spare time, Bradley continued working with his uncle on observations of Mars and Jupiter's satellites. By 1721 he was offered the Savilian Chair of Astronomy at Oxford and so he resigned from the church. Bradley spent the next few years working at observatories in Kew and Wanstead. There he noticed something strange, which he named "an aberration of light."

You've experienced an aberration without knowing it. When you're in a car or a train and it's raining, what happens to the rain? If the rain is falling straight down and you're not moving, then you'll see rain falling straight down, out of the side window. When you start to move, you are now moving toward the falling droplets, so from your perspective the rain now seems to be falling at an angle toward you a little, instead of straight down — the streaks of rain on the side window will be slanting toward you. The faster you move, the more the rain seems to be angled toward you. If you happen to be riding a bicycle in the rain, then you'll actually feel the effect. When you're stationary the rain will come straight down onto your rainhat. But when you start riding, you are riding into the falling droplets. From your perspective, the faster you ride, the more water you get in your face.

Above: Portrait of James Bradley, whose observations of the parallax phenomenon led to his discovery of the aberration of light.

Bradley believed he had observed the same effect happening to light falling on the orbiting Earth. This meant that light must be traveling at a finite speed, as Rømer had suggested. It also meant that if you knew how fast the Earth was orbiting the Sun, and you knew the angle that light appeared to fall on the Earth, you could work out how fast light actually traveled.

But spotting the aberration of light is not as easy as looking out of a train window on a rainy day. Light travels about 18 million times faster than rain falls. When you pass a station on a moving train, the light falling to the ground will also appear slightly angled toward you, but that angle is about 18 million times less than the angle of the rain (a tiny, tiny, tiny fraction of a degree), so you would never be able to see it. Amazingly, Bradley accidentally discovered the aberration of light through a search for a completely different effect, known as parallax.

You've also experienced parallax before, without knowing it. Imagine you're back on the train, looking out of the window. The world slides past you with a clackety-clack rhythm. The bridges and stations rush past in a blur. The more distant houses and trees go by more slowly. The really distant clouds in the sky move almost too slowly to see. You're experiencing parallax — the illusion of objects moving because of a change in the observer's position. None of the objects you were watching fly past you were moving. You were the one moving, in the train. But because the bridges are closer to you than the clouds, the bridges seem to move at a huge speed, while the clouds barely move.

Bradley had realized that as the Earth was orbiting around the Sun, it was moving at a tremendously fast speed relative to all the stars around us. He wondered whether the position of the nearer stars in the sky would appear to move more than the position of the stars that are further away. For several years he made meticulous measurements to see if he could detect the parallax of the stars. His results were initially very confusing. It appeared as though all of the stars were moving in little ellipses — which might mean that all the stars were the same distance from us! This was clearly wrong, and Bradley soon figured out the answer. The parallax of the stars was too small to detect (although later astronomers did detect it, and used it to figure out how far away the stars are). Bradley's strange movements of the stars were being caused by the aberration of light, and not the parallax of the stars.

Light aberration

The principle of light aberration is exactly the same as sitting on the train, watching the rain fall toward you. We're sitting on the Earth, which is moving around the Sun at a tremendous speed (roughly 67,000 miles an hour). If the Earth is at position E in space now, and at position E' in one hour, and we point our telescope at a star at position S, then we don't see the star where it really is. Because the light from S takes a few minutes to reach us, and because we're moving, the beam of light appears to be angled toward us. This means that we see the star at position S' and not its true position at S.

(If this is confusing, remember that every time you look into water, the refraction of light distorts the apparent position and size of submerged objects. In the same way, if you move fast enough, your speed will distort the apparent position of distant objects such as stars.)

So the line SE represents the light from the star S to the Earth E, with length determined by the speed of light. Our movement from E to E' causes the light beam to appear as though it follows the line $S'E$. The star will appear to be displaced from its true position by the angle SES', caused by the aberration of light. Because Bradley had made such good measurements, he could see the angles being formed, and he knew the speed the Earth was orbiting the Sun, so he was able to calculate the speed of light. His estimate was 301,000,000 meters per second — remarkably close to the value of 299,792,458 meters per second we recognize today.

Bradley went on to have a successful career in astronomy. He was appointed Astronomer Royal in 1742 and continued his research into the aberration of light. He also spent years building evidence that showed conclusively that the Earth nutates (its axis of rotation wobbles because of the gravitational pull of the Moon).

Subsequent astronomers found many new methods for measuring the speed of light, but it was another 200 years before Bradley's calculation was bettered. We now have many ways of measuring the speed of light, often using lasers. Astronauts have even put a mirror on the Moon, allowing us to flash a laser beam on it and measure the time taken for its light to return (Galileo would have loved the idea). Today the speed of light is treated as a definition because it is used to define the meter and other metric measurements (as we saw in Chapter 10). For now, the "official" speed of light, or c as it is known, will remain its current value whether anyone finds a more accurate way of measuring it or not.

Below: The aberration of light can be compared to the way that rain appears to fall at an angle as it streaks the windows of a moving train.

Above: Apollo *astronauts
deployed a Laser Ranging
Retro-Reflector in 1969 as part
of the* Apollo *lunar program.*

Seeing Is Not the Same as Hearing

Light has many other strange properties, which seem counterintuitive on first glance. One of these concerns relativity. (If this sounds like some kind of obscure physics, don't worry — it's not a difficult idea at all.) Imagine you're driving at 60 miles per hour, and someone driving the opposite direction passes you, also traveling at 60 miles per hour. From your perspective, the speed of the other driver relative to you is 120 miles per hour — that's why the other car flashes by so fast. Of course from the perspective of the driver of the other car, *you* are going 120 miles per hour relative to him, in the other direction — that's why you appear to speed past him so quickly. If the speed limit was 70 miles per hour, then relative to the other car you would have broken the limit. Relative to the road, you wouldn't have. Relativity is simply the idea that speed is *always* relative. (This is an old idea, which actually came from our friend Galileo.) The Earth is traveling at 67,000 miles per hour around the Sun, so relative to the Sun you are definitely breaking the speed limit, even when not driving anywhere. We typically use the convention of measuring speed relative to the Earth, but sometimes even this does not make sense.

Imagine the sound of the engines of those two cars. Sound travels at 742.3 miles per hour (more or less). The speed of your car can't make the external sound go any faster (it's not like throwing a ball ahead of you which will have the speed of your car plus the speed of your throw relative to your car, before wind resistance slows it). The speed of sound is caused by vibrating air molecules, which can't be pushed very effectively by your car. So if you're both going at 60 miles an hour, then the sound of your engines will easily race ahead of you and be audible to the other driver. But you've both spotted the red button inside the car — the one

that says, "Warning! Do not push!" You both push your buttons (despite the warning), and the two cars speed up to a terrifying 800 miles an hour. You are now driving faster than the speed of sound. Because sound does not move relative to your speed, it moves relative to the speed of the air surrounding Earth, you are driving faster than the sound of your engine. Anyone watching would see two rocket-powered cars flash past each other in total silence — and then would have the spooky experience of hearing the two deafening engines roar past each other, long after you were out of sight. (If you ever go to an airshow and watch the jetplanes flying, you'll have exactly this experience.)

Perhaps surprisingly, light in a vacuum behaves a little like sound in air. It doesn't matter how fast you go, light will always travel at the speed of light, and no faster. Riding in a spaceship traveling at half the speed of light and shining a light ahead of you will not "push" the light up to one and a half times the speed of light, any more than driving fast will push the sound faster.

But light is actually a bit stranger than sound. If I'm traveling in a train moving at half the

Above: A Navy Hornet fighter breaks the sound barrier. The pressure created by forward sound waves squeezes moisture in the air forming a ball of cloud over the front of the aircraft.

speed of sound, and I make a noise inside the train, then the noise is carried with me in the train. The noise travels at the speed of sound plus the speed of the train, which is one and a half times the speed of sound. So what about light? If I'm traveling in a spaceship at half the speed of light, then the lamps shining inside my spaceship must produce light traveling at the speed of light plus my own speed, which makes one and a half times the speed of light, right? Wrong. According to Albert Einstein, light can never travel any faster, even if the light source is moving, or you are moving relative to the source of the light. Now that is very weird indeed.

Special Relativity

Albert Einstein was born in 1879, in a town called Ulm, Germany. He was born to a non-practicing Jewish family and from birth was a little different from other children. His head was slightly larger than normal, and he spoke little, leading to one housekeeper to call him "retarded." At the age of 5, his father showed him a pocket compass. As an adult Einstein recalled this moment as being "one of the most revelatory events of his life," for as a child he had been astonished by the magical property of magnetism, which could travel through empty space and move the needle of the compass.

Einstein was taught in a Catholic school, but did not always do what he was told, believing that strict memorization was not a helpful way to learn. (As an adult, he wrote, "Education is that which remains when one has forgotten everything learned in school.") Luckily Einstein was able to interact with a medical student who came to visit every Thursday, and from Max Talmud (or Talmey as he was affectionately known), Einstein learned philosophy and mathematics, calling his copy of Euclid's

Elements the "holy little geometry book." Einstein also made models and mechanical devices as he grew up, and was assisted by his uncles who suggested key books on science and mathematics.

In 1894, when Einstein was 15, his father's business failed and his parents moved to Pavia in Italy. Einstein was left behind to complete his schooling. Rather than follow his parents' wishes, Einstein left the school early to join his parents. Nevertheless, he found the time to write his first scientific investigation (of magnetism) for his uncle, and at the age of 16 had made a breakthrough while looking in a mirror and wondering what he would see if he was traveling at the speed of light. The thought experiment became known as "Albert Einstein's mirror" and led Einstein to the belief that the speed of light is independent of the observer of the light. This idea was to become crucial later in his life.

Because he had not completed his schooling, Einstein's family sent him to Aarau, Switzerland, to obtain his secondary school certificate. It quickly became clear that he was not going to be an electrical engineer as his father hoped, with Einstein showing much more interest in studying electromagnetic theory and other aspects of theoretical physics. He graduated a year late at the age of 17, and entered the Swiss Federal Polytechnic Institute, moving to Zurich. The same year he renounced his citizenship of the German Empire and became stateless (not an official citizen of any country). By the time he was 23 Einstein had obtained a teaching diploma and also had a child out of wedlock with Mileva Maric, a medical student he had met at the Institute. The little girl was called Lieserl Einstein, but it is unknown whether she died in childhood or was adopted. Einstein married Mileva the following year. They went on to have

Above: From an unknown PhD student and clerk working in a patent office, Einstein changed the way the fundamental laws of the universe are understood.

Above: Nobel Prize winners Albert Einstein with Irving Langmuir, the American chemist (right), Sinclair Lewis and Frank B. Kellog at the 1933 Nobel Anniversary Dinner.

two sons, Hans Albert Einstein and Eduard Einstein. (As adults, Hans was to become a professor of hydraulics, but Eduard developed schizophrenia and died in an asylum.)

After graduating from the Swiss Federal Polytechnic Institute, Einstein struggled to find work. He was brash and seemed perhaps overconfident in his abilities, so was unable to obtain a position in any university. Instead, a friend helped him obtain an assistant position in a Swiss patent office. There he helped judge the technical feasibility of the ideas submitted to the office, and learned technical writing skills with the help of the director. He studied for his PhD at the same time, and in 1905 received his doctorate on "A new determination of molecular dimensions." But astonishingly, in the same year he also found the time to write four scientific papers that laid the foundation of modern physics. These became known as the *Annus Mirabilis Papers* (the *Year of Wonders Papers*). Three of those papers are widely considered to be worthy of a Nobel Prize: those on Brownian motion, the photoelectric effect and special relativity. The world of physics was stunned — here was an unknown PhD student and clerk in a patent office changing our understanding of the fundamental laws of the universe.

Special relativity still boggles the mind today. It's a very simple idea, based on two main

Special relativity

Time dilation (slowing) is predicted by special relativity. It's one way that we could make space travel more acceptable — for if we go fast enough, time actually slows down for us compared to the passage of time on Earth.

It sounds bizarre, but it's all because of that constant speed of light. Imagine you're on a train, and we've made a clock out of two mirrors and a light. Every tick of the clock happens when the light bounces from one of the mirrors. Because the distance between the mirrors is constant, and the speed of light is constant, this gives us a very accurate clock.

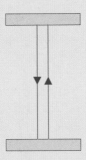

Now the train starts to move. Although the two mirrors are always the same distance apart on the train, they are now moving together, inside the train. If we were talking about sound this would be fine, for the air that transmits the sound moves with the train, so the sound would take exactly the same time to bounce between the mirrors. (To an outside observer, we would be speeding up sound by moving it with us on the train.) But remember that the speed of light cannot be altered. So as our train moves and the light bounces from one mirror, the second mirror is moving along, meaning the light has to travel a little further to catch up. When it bounces off that mirror, the first mirror is also moving along in the train, so again, the light has further to travel.

Even though the two mirrors are always the same distance apart, because we cannot "carry light" with us on the train, it's having to catch up with those moving mirrors. Because it's traveling a little farther, our clock ticks slower. The faster we move, the slower our clock ticks.

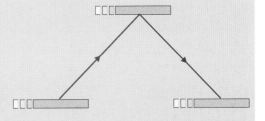

Special relativity is not some strange physicist's dream. In 1971, two scientists named Hafele and Keating synchronized several cesium atomic beam clocks (the most accurate clocks available at the time), then placed some on board normal commercial passenger jet planes, flying around the world twice. When they came back and compared the clocks that had been speeding inside the planes with those on the ground, they found that, as predicted, a time dilation effect had indeed made the clocks tell different times. Today, the GPS (global positioning system) which consists of many satellites traveling at high speed above us, only works because their clocks are adjusted for the relativistic time dilation effects.

All this means that if we could go fast enough inside a spacecraft (close to the speed of light), then time would slow down so much that we would barely age while decades flew by on Earth. It's a great way to reduce the time experienced by the passengers (but everyone on Earth would have a long, long wait to hear what happened).

Above: A German 55 euro cents special-edition stamp commemorates the 100th anniversary of the publication of Einstein's Theory of Relativity.

principles: first that the laws of physics will be exactly the same in all inertial frames of reference. This just means that the laws of physics "don't care" whether you drop a ball when sitting on a moving bus, or when sitting on the ground of the Earth as it orbits the Sun — the ball will fall from your hand in exactly the same way. It's basically saying, just as Galileo said, that all movement is relative, and so the resulting actions of forces will also be relative. The second principle is that c is invariant. Einstein was saying that the speed of light is independent of the motion of the light source, and independent of the observer of the light. So the speed of light is not relative. (That's why c is also called the universal constant — its value is not relative to anything else.) These two principles are important because they mean that all kinds of very strange things must happen in our universe. Special relativity tells us, for example, that time cannot be constant. Depending on how fast you move, you will experience time at a different rate (see box).

The other thing that falls out of special relativity is perhaps the most famous equation in the world: $E = mc^2$. While Einstein was not the first to spot this, his work was perhaps the most important explanation of why it was true. It's an equation that tells us why the speed of light is so important, for it says: energy equals mass multiplied by the speed of light squared. Put simply, you can transform energy into mass, or mass into energy — the two are equivalent. The equation also explains why an atomic bomb produces so much energy, because c squared is a very big number.

General Relativity

Einstein continued to work at the patent office, despite his scientific breakthroughs, until 1909, when he began to be recognized as a leading scientific thinker. By 1912 Einstein had secured himself the position of professor at the Swiss Federal Polytechnic Institute and he began to challenge astronomers to look for evidence of

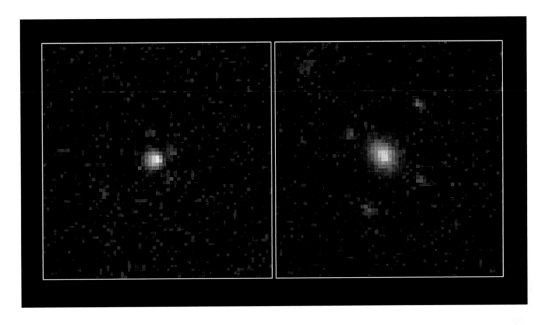

Above: Hubble images demonstrate gravitational lensing around a distant galaxy.

light being bent by gravitational fields of stars — something predicted by a new theory he was working on. Just before the beginning of World War I in 1914, at the age of 35, Einstein moved to Berlin and became the director of the Kaiser Wilhelm Institute for Physics. In 1915, he gave a series of lectures about his new theory, which he called general relativity. In his final lecture he presented a new equation to replace Newton's law of gravity, now known as the Einstein field equation. Once again, Einstein had astonished the scientific world. Instead of gravity being a force as Newton had described, Einstein explained gravity as a distortion in space-time caused by the presence of mass.

By 1919, Einstein's published work reached the ears of the newspapers. *The Times* published the story with the headline, "Revolution in science — New theory of the Universe — Newtonian ideas overthrown." The story was backed up by other scientists, who were quoted as saying that general relativity was "probably the greatest scientific discovery ever made," and the "greatest feat of human thinking about nature." Einstein became a celebrity overnight. The same year he divorced his wife and married his cousin Elsa (and heard the joke about "relativity" many times, no doubt).

Einstein received the Nobel Prize in 1921, but ironically it was for his work on the photoelectric effect, rather than his still-controversial theories of relativity. He spent the next few years traveling extensively, giving lectures around the world. In 1932 he was offered a part-time position in Princeton, to work five months a year. On his first visit there in 1933, Hitler came to power in Germany.

General relativity

As you may have guessed from the name, general relativity is a generalization of Einstein's earlier special relativity. The two principles of special relativity are still true, but Einstein wanted to explain gravity using the same ideas. From special relativity we know that energy, mass and light are linked (in the equation $e = mc^2$) and we also know that light and time are linked (for example in time dilation effects). General relativity basically says that mass and energy curve space and time.

In other words, despite what you may have been taught at school, Newton was not exactly right. Gravity is not a force. It's a field effect — a distortion caused by mass and energy. The easiest way to understand this is by imagining space and time is like a trampoline. If you put something with a big mass (like a heavy brick) on the trampoline, its surface is distorted and stretched. Now if you place a pool ball nearby on the trampoline, it will roll toward the heavy mass — like the gravity of a massive body pulling an object toward itself. If you flick the pool ball to one side, it will roll around the heavy mass, like a moon orbiting a planet.

But it gets stranger. Massive (and energetic) objects don't just create gravitational fields in space, they also distort time. The more massive the object, the slower time goes by. The further away from the mass you go, the faster time goes. Once again, these effects have been measured — it is even possible to detect the tiny, tiny difference in time at the top of a tall building compared to at the bottom (at the top you're a little further from the mass of the Earth). It's also quite easy to see the "gravitational lensing" effect predicted by Einstein of distant stars and planets that bend the path of light around them because of the distortions in space-time they produce.

When space goes bendy, it means that Euclidian geometry doesn't work so well anymore. Near to a massive object like a black hole, the angles of triangles will not all add up to 180 degrees any more (in the next chapter we'll see more of black holes). Luckily there are no black holes near to us. Like Newton's equations, Euclid's geometry is right most of the time, but Einstein is right even more of the time.

The general theory of relativity also makes it easier to understand why we can never go faster than the speed of light. The equations tell us that we need more and more energy to produce the same acceleration (or the same energy produces less and less acceleration, the faster we go). If you wanted to travel at the speed of light, you would need an infinite amount of energy. That's why we can't do it. That's also why traveling faster than the speed of light (as they pretend to do in science fiction) is nothing but a fantasy. You can't generate more than an infinite amount of energy. It may be that the only plausible way to travel long distances in space and bypass the speed limit of c would be to punch through a wormhole (as we saw in Chapter 3).

*Above: Harold Lyons explains
the experiment in which an
"atomic clock" is mounted in
an orbiting satellite to check
Einstein's theory of relativity.*

Before long, the "Law of the Restoration of the Civil Service" was passed in Germany, forcing all Jewish professors to lose their jobs.
A campaign began in Germany to discredit Einstein's work and call it "Jewish Physics," in contrast to "Aryan Physics." Einstein never returned to Germany.

Not merely content with general relativity, Einstein spent the rest of his life attempting to unify the laws of physics, especially gravitation and electromagnetism. He wanted to find an equation as elegant and simple as $E = mc^2$ that explained magnetism, electromagnetic waves, gravity and everything else in physics. He was not successful in his attempts, saying "I have locked myself into quite hopeless scientific problems — the more so since, as an elderly man, I have remained estranged from the society here ..." The goal of explaining everything by a single, elegant, unified equation is still pursued by physicists today. Anyone who manages it will make a breakthrough as big as Einstein's. No one has managed it yet.

In 1939 Einstein was persuaded to write to President Roosevelt urging that a research program should be started on nuclear fission, in case the Germans developed nuclear weapons first. The president began a small investigation, which before long became the Manhattan Project, which resulted in the nuclear weapons used at the end of World War II. Einstein later regretted writing the letter.
In 1940 he became a citizen of the United States, and helped the war effort in 1944 by handwriting his 1905 paper on special relativity, and selling it at auction for $6 million. (Today it is kept in the Library of Congress.)

Einstein may have always felt an outsider with very different goals to those of most people. As an old man Einstein wrote, "I have never belonged wholeheartedly to a country, a state, nor to a circle of friends, nor even to my own family. When I was still a rather precocious young man, I already realized most vividly the futility of the hopes and aspirations that most men pursue throughout their lives. Well-being and happiness never appeared to me as an absolute aim. I am even inclined to compare such moral aims to the ambitions of a pig."

Einstein began to suffer from ill health in 1950. In 1952 he was rather embarrassed by an offer from the Israeli government for the job of second president. He politely refused. One week before he died in 1955, Einstein signed his name for the last time on a letter to Bertrand Russell, agreeing for his name to be placed on a manifesto to urge all nations to give up nuclear weapons.

During his lifetime, Einstein had transformed our ideas of the universe and existence. Perhaps his time on our small planet can be summarized best in his own words: "Each of us visits that Earth involuntarily and without an invitation. For me, it is enough to wonder at its secrets."

Above: Crown of a nuclear bomb exploding in French Polynesia.

How big can a number be? One of the largest numbers we have a name for is centillion. If you wrote it down, it would be a 1 followed by 600 zeros. But that's not the biggest number. Googolplex is much bigger – you'd have to write a 1 followed by googol zeros (where a googol is itself written as 1 followed by 100 zeros). And there are bigger numbers than this. Some are so absurdly huge that they can't be written down without inventing new notation, such as Graham's Number and Moser, named after their creators.

THE NEVERENDING

CHAPTER ∞

Big numbers are actually very easy to create. For example I could create my own Bentley's Number, and ensure that it's the biggest so far by multiplying centillion by googolplex by Graham's Number by Moser, and then multiplying that by the product of every number in this book, including the page numbers. As big as it sounds, it's hardly the biggest number possible. Just imagine the number you'd get if you multiplied the Bentley's Number by itself. Yet, the result of that product is still minuscule compared to the number you'd get if you raised that number to the power of itself. And the result of that exponential would be invisibly small compared to the number you'd obtain if you raised it to the power of itself again.

STORY

Numbers may provide us with a language for describing the universe, but they are not limited as our universe is. We can use numbers so big that they surpass the number of everything in the universe. (The number of atoms in the universe is tiny compared to the numbers we were just looking at.) There is no limit to the size of the numbers we can use. Our numbers can be as big as we want them to be. One of the cleverest things about numbers is that there is no biggest number. (If you think you've found it, I'll just add 1 to it and have a bigger number than you.)

Sometimes, however, we want to think about a different concept. Not a number, but an idea. It's the idea of neverending, never stopping, going on forever, eternity. We call that idea infinity.

The Beginning of Forever

Philosophers have worried about infinity for thousands of years. With a seemingly infinite universe stretching from us in all directions, it's not hard to see why. If you have no science to help explain the extents of the universe, then you must rely on arguments that seem to make sense. If space does not go on forever, then what does its edge look like? What would happen if you threw an object off the edge of the universe? Where would it go? How could there even be an edge without something on the other side?

From arguments like these, it seemed clear to some philosophers that space goes on forever. But not everyone agreed. Even the concept of "forever" seemed difficult to reconcile with our world around us. From observation, nothing seems to last forever. Could anything be infinitely big? Or even infinitely small?

One of the earliest people to investigate these ideas was a man called Zeno, born about 490 B.C.

Zeno.

Left: Zeno, who examined the
idea of divisions into infinitely
smaller quantities to prove the
truth in monism. His ideas were
discredited by Aristotle.

in Elea, Lucania (today Southern Italy). Zeno was
a student of a philosopher called Parmenides at
the Eleatic School, one of the leading schools of
Greek philosophy begun by Parmenides. Zeno
was taught monism: a belief that all things are
aspects of a single eternal reality called Being.
Change was impossible – in monism all was one,
and non-Being was inconceivable.

Zeno found such ideas intriguing and wrote a
book of paradoxes that were to puzzle and frustrate
philosophers for centuries to come. According to
Plato, the book was stolen and published (which
in those days meant, "carefully copied by hand")
without Zeno's permission. Nevertheless, Zeno
became well-known for his work and later visited
Athens and met the young Socrates.

Zeno's paradoxes were mainly intended to
affirm that everything was one, as monism
taught. To do this, he considered what would
happen if you tried to divide something into

Zeno's Achilles paradox

The paradox works like this. Imagine Achilles is
a fast runner who has decided to race a slow
tortoise. He lets the tortoise have a head start
as it runs so much slower than him. After the
tortoise has moved 100 meters from the start,
Achilles sets off. It doesn't take him too long to
reach the point where the tortoise had been.
But in that time, the tortoise kept on running,
so now it is 5 meters away. Achilles keeps
running, and soon reaches the point where the
tortoise had been again, but the tortoise is still
moving, and is still a small distance (25 cm)
ahead. Achilles is still running and so in
almost no time he reaches the point where the
tortoise had been, but it also kept moving, so
now it is 1.25 cm ahead. In the tiny fraction of
time it takes Achilles to reach that position, the
tortoise has moved again, and is now less
than 1 mm ahead. And so it goes on. Every
time Achilles reaches the position where the
tortoise had just been, the tortoise has already
moved just a tiny bit. Somehow, the fast
runner can never overtake the tortoise!

infinitely small amounts, showing that when-ever you tried, the result somehow made no sense at all. One of his most famous paradoxes was about a runner called Achilles. Rather astonishingly, Zeno managed to construct a seemingly logical argument that shows that Achilles would never be able to overtake a tortoise running slower than him.

Aristotle wasn't very impressed with Zeno's arguments, calling them fallacies without really being able to dispute them. (It was 2,000 years before we had the mathematics to explain how even with such infinitely decreasing series, Achilles would win the race. The paradox works because we're focusing our attention on an ever-smaller period of time and space just before Achilles reaches the tortoise. If we watched him, he'd run straight past without a pause.) Zeno had used numbers that shrank to become infinitely small in order to make his apparent paradox, but Aristotle's view of infinity was altogether more practical.

Aristotle was born in 384 B.C., in Stagirus, Macedonia, Northern Greece. His father, Nicomachus, was a medical doctor, so it is likely that Aristotle's early years were spent accompanying him on visits to tend to patients. But his father died when Aristotle was only 10, so instead of becoming a doctor himself, Aristotle was raised by his uncle and was taught Greek, rhetoric and poetry. When he was 17, he

Above: Aristotle, whose belief in a finite universe is one that is ultimately upheld by the church.

joined Plato's Academy in Athens. (If you recall, we met Plato in Chapter √2. Plato was the one who liked to write in riddles.) Aristotle stayed at the Academy for 20 years, first as a student, then as a teacher. He left at around the time of Plato's death, perhaps because he disapproved of the new leader Speusippus.

Aristotle moved to Assos, which faces the island of Lesbos, and with the support of the ruler Hermias of Atarneus, led a group of

Above: Manuscript illumination of Aristotle tutoring Alexander the Great from The Adventures of Aristotle.

philosophers. He began to develop his own beliefs that were distinct from Plato's teachings, with a strong focus on biological observation and anatomy. He also married the niece of Hermias and had a daughter, Pythias. Sadly his wife died only 10 years later. Political unrest forced Aristotle to move once again (the Persians attacked the town and executed Hermias). Before long, Alexander (eventually known as Alexander the Great) came to power.

As well as supporting the Academy, Alexander asked Aristotle to start a second place of learning, which Aristotle named the Lyceum. Unlike the Academy, which taught a very narrow range of topics, Aristotle encouraged a much broader education. He gave many lectures himself on topics including logic, physics, astronomy, meteorology, zoology, metaphysics, theology, psychology, politics, economics, ethics, rhetoric and poetics. In doing so, Aristotle helped invent many of these areas, for some had never been formally taught before. His ideas were so persuasive and influential that many of the major philosophical and scientific ideas in the Western world became Aristotelian for the next 2,000 years. Through observation by eye and logical thinking, Aristotle had produced many explanations of how the universe worked. Not all of them were correct; as we've seen in earlier chapters, he described the geocentric view of the universe (the Earth in the middle, and everything orbiting around it) and believed that light was instantaneous. He also considered infinity.

Aristotle couldn't deny the evidence he believed he saw for infinity. He could not conceive of a beginning or end to time, so he felt that the abstract notion of infinity could

Above: Theologian Thomas Aquinas' writings of the 13th century were a conduit for Aristotelian philosophy.

exist. But he viewed it more as a potential than a reality. Aristotle gave an example: imagine you are describing the Olympic Games to someone. All you can do is describe them as a potential – they should happen in the future, but are not happening right now. The same is true of infinity, according to Aristotle. It has the potential to exist, but doesn't exist now (and may never in the future). Nothing in our physical world could have infinite size or age, so you'll never come face-to-face with anything infinite.

These views of infinity became the accepted wisdom and most of mathematics operated very nicely for centuries. Through the writings of theologian Thomas Aquinas in the 13th century, Aristotelian philosophy became from

Above: Giordano Bruno, Italian author of philosophical dialogues and works on mathematics and physics.

then on intricately linked to religion. God was considered to be infinite (you will never meet God in our physical world, for example). The notion of an everlasting, infinite soul became integrated into religion — a comforting thought that contrasted well with the bleak prospect of a neverending nothingness of non-existence for the non-believer. The church committed itself to the Aristotelian view that our universe is finite, with the Earth at its center. (In some texts, measurements for the dimensions of heaven can be derived — and according to those, heaven was not very big.) The notion that the points of light in the sky might be distant suns, and that the universe might be infinite was not just seen as ridiculous, it was heresy. Those who dared oppose the view of the church were not tolerated well. Giordano Bruno was not a mathematician or a scientist, but he nevertheless wrote a book entitled *On the Infinite Universe and Worlds* (1584). He was tortured for nine years by the Inquisition in an attempt to force him to revoke his belief in the infinite. Bruno never gave in — indeed he may have even deliberately challenged the Inquisition on the subject. Eventually in 1600, he was gagged to prevent onlookers hearing his views, then burned at the stake.

Ironically, modern science agrees with Aristotle and the church — the universe is very, very large, but it is thought to be finite, not infinite.

Wheels within Wheels

Galileo was only too aware of the fate of Bruno. They lived in dangerous times to be challenging the orthodoxy. Yet, that didn't stop the imaginative Galileo from thinking about the world. According to the Aristotelian view, infinity was only a potential, never a physical reality. Galileo couldn't help but notice something a little strange about circles. He considered two circles, one larger than the other, and tried to understand how many points there might be on the circumference of those circles. He came to the conclusion that there must be an infinite number of points on both, even though one was bigger than the other (see box)!

Galileo was troubled by this, for it did not make any sense. How could one infinity be larger than another? Surely infinity was, by definition, infinite? He wrote, "... we attempt, with our finite minds, to discuss the infinite, assigning to it properties which we give to the finite and limited; but I think this is wrong, for we cannot speak of infinite quantities as being the one greater or less than or equal to another."

He didn't make it easy for himself though. Galileo next thought about the list of all positive integers, and the list of all possible squares of integers. For every possible integer there is a single square: 1:1, 2:4, 3:9, 4:16, 5:25, and so on.

Galileo's Circles

Galileo thought about two concentric circles (one inside the other with their centers in the same place). One clearly had a longer circumference than the other.

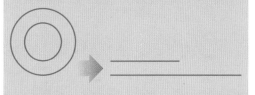

Then he imagined what would happen if a line swept around the circles, like the hand of a clock.

At any point in time, that line must be crossing the larger circle at exactly one point. But it must also be crossing the smaller circle at one point as well! Even though the larger circle is longer and so must have more points on it, for every specific point on the larger circle's circumference, there must be a matching point on the smaller circle's circumference.

Even if the line is swept around the circles by an infinitely small amount, it will always cross the larger circle at one point and the smaller circle at one point. So there must be an infinite number of points making up the smaller circle and a larger infinite number of points making up the larger circle.

This must mean that there are the same number of squares as there are integers. Yet it is clear that many numbers are not squares, so there must be more integers than squares. Somehow the number of integers is both more than and the same as the number of squares. But both lists go on forever, so both are infinite. To Galileo, the solution was that,

> … the totality of all numbers is infinite, and that the number of squares is infinite; neither is the number of squares less than the totality of all numbers, nor the latter

Below: Georg Cantor is best known as the creator of set theory, which has become a foundational theory in mathematics.

greater than the former; and, finally, the attributes "equal," "greater" and "less" are not applicable to the infinite, but only to finite quantities.

We had to wait for the quirky genius of Georg Cantor to understand these strange aspects of infinity. (If you recall, we met Cantor and saw his obsession with Shakespeare and Francis Bacon in Chapter √2.) Cantor noticed a property of numbers similar to the one that had confused Galileo. He decided to try and prove that some sets of numbers cannot be paired with the integers as Galileo had done with squares. If there is a set of numbers that cannot be matched one-to-one with the integers, then it must be bigger than the set of integers — even though both sets of numbers are infinite (see box opposite).

Cantor had shown that some sets were uncountable and, even more amazingly, some infinite sets are bigger than others. Infinity was no longer simply "as big as you can get." Now we understood that every infinity might be bigger or smaller than every other infinity. Just because infinity goes on forever doesn't mean it's always the same size.

Meeting the Infinite

Galileo and Cantor lived in different worlds 300 years apart, so Cantor could never explain to Galileo that "greater" and "less than" could be used with the infinite. But what of Aristotle's 2,000-year-old view that infinity is only a potential, and never a reality? Certainly we will never be able to write down Cantor's infinite

Cantor's diagonal argument

Cantor proved that some infinite sets are bigger than others using several clever methods. One of the most famous has become known as Cantor's diagonal argument.

Cantor decided to create an infinite set of numbers. This would be an infinitely big set of infinite lists of 1s and 0s, each list in some kind of neverending pattern:

{0, 1, 0, 1, 0, 1, 0, 1,... }
{1, 1, 0, 0, 1, 1, 0, 0,... }
{0, 0, 1, 0, 0, 1, 0, 0,... }
....

Next he imagined constructing a new infinitely long list of numbers, constructed from this infinite set. The new list would have its first number different from the first number of the first list in the set. It would have its second number different from the second number of the second list in the set, the third number different from the third number of the third list in the set, and so on, forever.

{0, 1, 0, 1, 0, 1, 0, 1,... }
{1, 1, 0, 0, 1, 1, 0, 0,... }
{0, 0, 1, 0, 0, 1, 0, 0,... }
New list: {1, 0, 0,... }

Cantor then explained that the new list of numbers could not appear in the infinitely long set. Whichever list of numbers in the set he compared the new list with, because of the way it is constructed, it will be different. For example, if he chose the 100th list in the set, the new list would be different at its 100th number. If he chose the 3,333rd list in the set, the new list would be different at its 3,333rd number. Cantor's diagonal argument (named because of the diagonal pattern of numbers used to make the new list) proves that an infinite set of lists of numbers cannot contain all possible lists of numbers. In other words, there are more sets of numbers than there are numbers. Not only that, but there are more sets of sets of numbers than there are sets of numbers. And there are more sets of sets of sets of numbers than there are sets of sets of numbers, and so on ...

lists, nor will we be able to plot Galileo's infinite points on his circles. Was Aristotle correct? Is there nothing truly infinite in our universe?

The answer to that depends on how clever Einstein was. According to Einstein's theory of general relativity, an object massive enough will distort space and time so much that its gravitational field will collapse its own structure. (Imagine the Earth being as fragile as paper while keeping the same mass — its gravity would pull its own surface inward, crumpling it like a paper ball.) Einstein's equations predict that a star that is massive enough may do the same thing, squishing itself into a smaller and smaller volume. A physicist named Karl Schwarzschild used Einstein's field equations to calculate what we now call the Schwarzschild radius for any mass. This is the size at which a non-rotating object collapses completely under its own gravity.

Above: Karl Schwarzschild investigated the effects of gravity and its relationship to objects that have zero size and infinite mass.

If our sun was squished to a ball with a radius of 2 miles (3 km), or if the Earth was squished to the size of a marble with radius of 0.35 inch (9 mm), then they would become black holes. Everything would be sucked into them by their enormous gravitational fields, even light (which is why they're called black holes). They would continue to pull their own surfaces toward their centers until they became singularities: points with zero size and infinite mass. Some suggest that a rotating black hole might create a ring-shaped singularity that would then act like a wormhole in space-time, punching through the universe to connect with another rotating black hole somewhere else in the universe (remember the hole punch in the Möbius strip in Chapter 3). But the science fiction stories that rely on travel through wormholes neglect to mention the horrific gravitational effects that would tear you to pieces smaller than atoms long before you were even very close, not to mention the eon-length time dilation effects.

The origin of the universe is also predicted by general relativity to be a "causal singularity." That's a point of no size or time which exploded in the Big Bang, creating everything in our universe, including space and time.

We even have evidence of these mind-boggling concepts. The Hubble Space Telescope has photographed several images of huge rotating gas clouds at the heart of distant galaxies. Because they rotate, the radius and speed of the constituents can be measured, and this enables the calculation of their mass. Astronomers can show that these distant objects are monstrously massive, and so

because of their size these must be black holes, busily sucking up the gases and stars around them.

We also have plenty of evidence for the Big Bang that began our universe. We've known that the universe is expanding since Edwin Hubble discovered in 1929 that all the galaxies are moving away from a single point. In fact the galaxies are not moving, but the very space

Above: The Hubble telescope has revealed vast rotating gas clouds at the center of galaxies, which are thought to indicate the presence of black holes.

they're in is moving. Draw lots of dots on a balloon, inflate that balloon and you'll see the same effect — the dots appear to move away from each other, but it's the surface they sit on that is expanding. (This is why we know the universe is a finite size — it began as a point and it has only had enough time to expand to a certain size so far.) We can also measure the background heat in the universe that is left over from the high temperatures in that initial explosion. The calculations tell us that our universe is between 13 billion and 14 billion years old. To provide some comparison, the Earth and our solar system is about 4.54 billion years old. From fossil evidence we know that

Above: Illustration of a star field with a circular black hole at its center.

Opposite: Studying the formation of stars has helped us learn about the Big Bang itself, and ultimately led to the conclusion the universe is finite.

primitive life arose on Earth around 3.5 billion years ago.

So does all this evidence mean that there are singularities in the universe? Points with infinite mass and no size? The short answer is, we don't know. We can't actually go and have a close look at a black hole or the Big Bang. Einstein's equations predict singularities, but other equations do not. So far in physics, whenever infinity appears in an equation it has eventually turned out that the equation was slightly wrong, and many physicists believe this is true for the Big Bang and black holes. It's why there is still much to do in physics to understand our universe. We've come a long way with the help of some stunningly clever

people, but we need new geniuses to help figure it all out. The great 20th-century physicist Richard Feynman once spoke about infinity, saying, "It is my task to convince you not to turn away because you don't understand it. You see, my physics students don't understand it either. That is because I don't understand it. Nobody does."

Maybe Aristotle was right all along. Maybe infinity is something we can conceive of, but never see. Unless you think you know better?

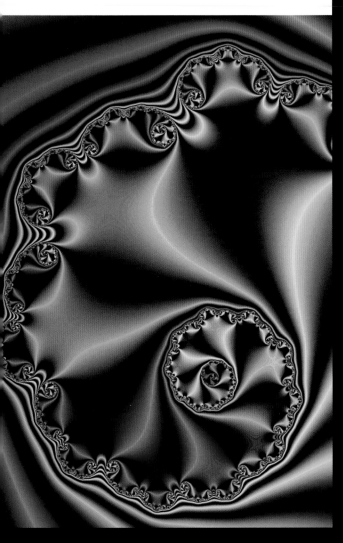

The rich tapestry of numbers that forms our history and future are all closely linked to reality. To make a fraction, just cut an apple into quarters; to see pi, look at a circle; to see anything at all, open your eyes and experience photons hitting your retinas at the speed of c. All numbers are entwined with our physical world in this way. But there is one number that has a more complicated relationship with reality. It's known as a complex number, or an imaginary number, or more often simply: i. Despite these names, it's very real indeed.

UNIMAGINABLE

CHAPTER i

The imaginary number is the answer to a puzzle that perplexed mathematicians for centuries. The puzzle emerges when thinking about squares and square roots. As we saw in Chapter $\sqrt{2}$, the square root function simply tells us what number we should multiply by itself (square) to produce the current number. Or to put it another way, square root tells us the root of the square. The square root of 2 must be more than 1 (because $1 \times 1 = 1$) and less than 2 (because $2 \times 2 = 4$). It turns out that $\sqrt{2}$ is the irrational number: 1.414213562373095… If you wanted to, you could work that out by creating a very accurate square with sides 1 foot and measuring the distance from corner to corner across the diagonal.

Square root is a very easy idea, then. So, the puzzle is, what's the square root of -1? What is the value of $\sqrt{-1}$?

The answer can't be -1 because $-1 \times -1 = 1$. Equally, the answer can't be 1, because $1 \times 1 = 1$ as well. You can't draw a square with sides of -1 and if you're trying to find the answer from your calculator, you won't receive much except an error — your calculator doesn't know the answer to this puzzle either.

Using Your Imagination

The puzzle has been known for centuries. As long ago as A.D. 50, the Greek mathematician Heron of Alexandria discovered it when trying to calculate the volume of part of a pyramid. The first person to actually use the square root of a negative number in his mathematics was an Italian named Niccolo Fontana.

Fontana was born in Brescia, near Venice in 1500. He was the son of a mail rider (who used a horse to deliver mail to neighboring towns). At the age of 6, his father was murdered, plunging his family into poverty. Fontana's life did not become any easier. At the age of 13 his

COMPLEXITY

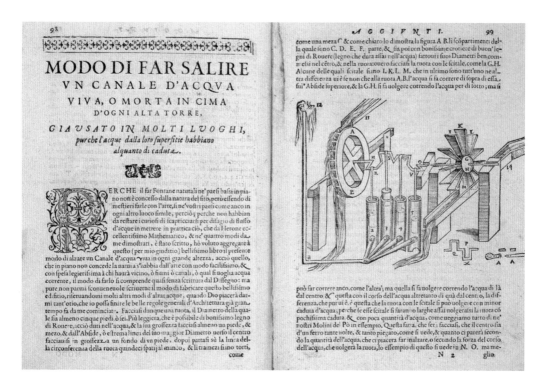

town was invaded by the French army, killing around 46,000 residents of Brescia. Fontana hid in a nearby cathedral with his sister, but soldiers found them and used a saber to slash his face, injuring him terribly. He was left for dead, but with the careful nursing of his mother (who sadly could not afford any medical help), he recovered. For the rest of his life he bore terrible scars on his face and was unable to speak normally, leading to his nickname of *Tartaglia*, or stammerer. As an adult he grew a beard to cover his scars.

Fontana taught himself mathematics, and while he showed remarkable promise, his overconfident and cocksure demeanor seems to have made it difficult for him to find work. By the age of 18 he had married and started a family, while teaching at a school. At the age of 35 he moved to Venice for a better paid job as a mathematics teacher in another school.

Despite being a stuttering school teacher (or perhaps because of it), Fontana soon developed a reputation as an impressive mathematician,

Cubic equations

Cubic equations are equations that have one term in them that is to the power of 3 — something cubed. So a typical cubic equation might look like this:

$$x^3 - 6x^2 + 11x - 6 = 0$$

which if we plotted, describes a curve like this:

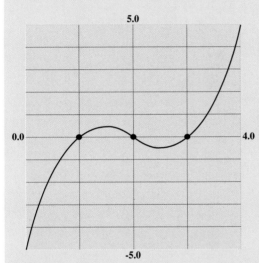

This curve crosses the x-axis in three places, at (1, 0), (2, 0) and (3, 0). These are called the solutions to the equation. We know the solutions are right, even without drawing the graph because if we put the numbers into the equation, the answer is 0 each time:

$x = 1$ 1 x 1 x 1 − 6 x 1 x 1 + 11 x 1 − 6 = 0

$x = 2$ 2 x 2 x 2 − 6 x 2 x 2 + 11 x 2 − 6 = 0

$x = 3$ 3 x 3 x 3 − 6 x 3 x 3 + 11 x 3 − 6 = 0

But if we put any other number into the equation, the answer is not 0. (Try, if you don't believe me.)

The problem faced by mathematicians was: how can you figure out the solutions of any cubic without drawing it? It's a tricky problem, because sometimes the "wiggle" of the curve is a very different shape, or above or below the x-axis, so there are not always three solutions.

skilled at performing in mathematics debates. These were like sports competitions, often open to the public, at which two mathematicians would challenge each other with the latest and most difficult mathematics problems, the winner being the one who could solve the most problems.

Fontana wrote several important books, and translated Euclid's *Elements* into Italian and Archimedes' works into Latin for the first time. But it was his work on cubic equations that eventually was to ruin him.

Another Italian mathematician called Scipione del Ferro had discovered a partial solution for solving cubic equations — he could solve simple ones that looked like: $x^3 + ax = b$. He kept it a secret, only telling his assistant Fior on his deathbed. Several years later, Fior began to

QVESITI ET INVEN-
TIONI DIVERSE
DE NICOLO TARTAGLIA,

DI NOVO RESTAMPATI CON VNA
GIONTA AL SESTO LIBRO, NELLA
quale ſi moſtra duoi modi di redur una Città ineſpugnabile.

LA DIVISIONE ET CONTINENTIA DI TVTTA
l'opra nel ſeguente foglio ſi trouara notata,

CON PRIVILEGIO

APPRESSO DE L'AVTTORE
M D L I I I I.

Above: Title page from Tartalias
General tratto di numeri et
misure *(1556).*

Fontana was clever, and sent Fior a range of different types of problems. But Fior believed that only he could solve cubic equations, so sent 30 cubic problems. Fontana solved them all in two hours, demonstrating his superior mathematics skills to the world.

Others soon learned of Fontana's success and were eager to learn what his general solution to the cubic might be. Fontana was initially very reluctant to tell anyone, but tempted at the prospect of his discovery leading to a better job, he relented, providing his method to a famous doctor and mathematician called Cardan. He wrote it down in the form of a poem in an attempt to make it more difficult to be stolen by a rival mathematician. Cardan promised never to publish his secret. Brace yourself — the poem is perhaps not to the taste of lovers of conventional poetry (maybe it was better in Italian):

When the cube and things together
Are equal to some discreet number,
Find two other numbers differing in this one.
Then you will keep this as a habit
That their product should always be equal
Exactly to the cube of a third of the things.
The remainder then as a general rule
Of their cube roots subtracted
Will be equal to your principal thing
In the second of these acts,
When the cube remains alone,
You will observe these other agreements:
You will at once divide the number into two parts
So that the one times the other produces clearly

boast to other mathematicians that he could solve any cubic equation. Since Fontana had already figured out a solution to cubic equations that looked like this: $x^3 + ax^2 = b$, he challenged Fior to a contest. Each mathematician would send each other 30 of their hardest problems, and the winner would be the one with the most answers solved in the shortest time.

The cube of the third of the things exactly.
Then of these two parts, as a habitual rule,
You will take the cube roots added together,
And this sum will be your thought.
The third of these calculations of ours
Is solved with the second if you take
good care,
As in their nature they are almost matched.
These things I found, and not with sluggish
steps,
In the year one thousand five hundred, four
and thirty.
With foundations strong and sturdy
In the city girdled by the sea.

Cardan noticed early on that Fontana's method led to some strange effects with numbers. He discovered that when attempting to find the solution

Below: Portrait of Gerolamo Cardan.

Cardan's puzzle

To see how Fontana and Cardan used imaginary numbers, take a look at a puzzle tackled by Cardan:

Divide 10 into two equal parts, the product of which is 40.

In Cardan's own words, "It is clear that this case is impossible. Nevertheless, we shall work thus: we divide 10 into two equal parts, making each 5. These we square, making 25. Subtract 40, if you will, from the 25 thus produced,… leaving a remainder of -15, the square root of which added to or subtracted from 5 gives parts the product of which is 40. These will be $5 + \sqrt{-15}$ and $5 - \sqrt{-15}$.

Dismissing mental tortures, and multiplying $5 + \sqrt{-15}$ by $5 - \sqrt{-15}$, we obtain $25 - (-15)$. Therefore the product is 40 … and thus far does arithmetical subtlety go, of which this, the extreme, is, as I have said, so subtle that it is useless."

Amazingly, for this problem Cardan had found a way to divide 10 into two equal numbers, both comprising a real part (5) and an imaginary part ($\sqrt{-15}$). Any square root may be either positive or negative (for example, the square root of 4 is both 2 and -2, because $2 \times 2 = 4$ and $-2 \times -2 = 4$). So Cardan made his two equal parts be the real number plus the square root, and the real number minus the square root. When added they equal 10; when multiplied they equal 40.

*Above: Painting of Venice, "the
city girdled by the sea."*

of some cubic equations, the results involved taking
the square root of negative numbers. He wrote to
Fontana, trying to obtain help, but Fontana now
regretted telling Cardan his secret so he wrote a
cryptic reply, deliberately trying to confuse Cardan:

> and thus I say in reply that you have not
> mastered the true way of solving problems
> of this kind, and indeed I would say that
> your methods are totally false.

But Cardan was correct, and soon managed
to figure out how Fontana could solve equations
even when a negative square root did appear.
He decided to treat such results as though they
were numbers.

Cardan quickly developed Fontana's method for solving cubics and with the help of his assistant Ferrari even managed to extend it to solve quartic equations. They also discovered that the original inventor of a solution to cubic equations had been del Ferro, so they published del Ferro's method and their own work, bypassing the promise to Fontana that they would keep his method secret forever.

Fontana was furious, and published his work in a book, the following year. As well as writing the story of his discovery, he also added several insults and malicious comments about Cardan. Ferrari learned of this and began a heated exchange by letter with Fontana. In one letter, Ferrari wrote to Fontana:

> You have the infamy to say that Cardano is ignorant in mathematics, and you call him uncultured and simple-minded, a man of low standing and coarse talk and other similar offending words too tedious to repeat. Since his excellency is prevented by the rank he holds, and because this matter concerns me personally since I am his creature, I have taken it upon myself to make known publicly your deceit and malice.

Like a duel by numbers, Ferrari challenged Fontana to a public debate. Fontana refused, wishing only to challenge Cardan. But in 1548, Fontana was offered a prestigious lectureship in Brescia. To prove his worth as a mathematician, Fontana agreed to the debate with Ferrari. When the day came, Fontana discovered to his horror that Ferrari's extensive work on cubics and quartic equations meant that Ferrari understood them much better than he did. Rather than risk the embarrassment of losing, Fontana quietly slipped away at night after only one day of the two-day event. Ferrari was declared the winner, by default.

Fontana's forfeited debate was a mistake. After lecturing for a year in Brescia, he discovered that he would not be paid for his work. Despite taking legal action, he was forced to return to his previous job. While Cardan went on to become one of the most famous (and controversial) doctors and mathematicians of his era, Fontana died in poverty at the age of 57, in Venice.

Drawing on Imagination

From the time of Fontana, imaginary numbers continued to appear regularly in mathematics. There was still considerable confusion about them and dislike for them. Descartes (the inventor of Cartesian geometry) coined the name "imaginary" for the numbers, using the word in a derogatory sense — it was surely so much better to use real numbers. A few decades later, mathematicians de Moivre and our argumentative friend Newton combined trigonometry with complex numbers to solve some of the trickier problems Cardan had attempted. Still later, Euler (the inventor of most of our modern mathematical notation, who lost his sight toward the end of his life) gave us i. It was a simpler way of writing down the imaginary number $\sqrt{-1}$ which hid the specter of a negative square root from sight, and has been used ever since.

Three hundred years after Fontana was born, Norwegian surveyor and map-designer Caspar Wessel was the first to draw imaginary numbers geometrically. Sadly, Wessel's work lay

Argand diagrams

Argand diagrams are very simple geometric diagrams that allow us to visualize imaginary numbers. Each imaginary number is written as a pair of numbers: $a + bi$, with a representing the real part and b defining the size of the imaginary part. An imaginary number is drawn on a pair of axis, where the x-axis (the horizontal one) defines the real part, and the y-axis (the vertical one) defines the imaginary part. So to draw the imaginary number: $2 + 3i$ is very easy:

imaginary numbers — which is exactly what Wessel and Argand suggested. For example, if you wanted to add two imaginary numbers together, just plot them on an Argand diagram, draw a line from the origin (0,0) to each one, to form the vectors $z1$ and $z2$, and the sum of the vectors gives a vector to the new imaginary number.

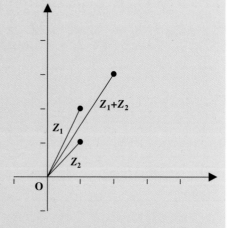

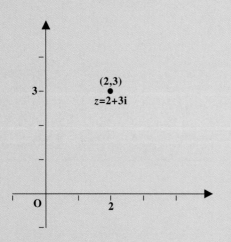

If you can draw a single number as a point, then you can use trigonometry and manipulation of vectors to calculate using

When subtracting you take one vector from the other, when multiplying you add the angles between the vectors and the real axis then multiply the length of the vectors, and so on.

unnoticed for 100 years, and it was a Parisian bookkeeper named Jean-Robert Argand who was to take all the credit. Rather unfairly, geometric diagrams of complex numbers are today called Argand diagrams, not Wessel diagrams (see box, opposite).

Turning Dreams into Reality (or Reality into Dreams?)

Imaginary numbers remained a mathematical enigma for decades. While we could imagine them and even draw them, it was hard to understand what they really meant. If pi corresponded to circles, then did i correspond to a previously unknown and mysterious something in nature? If i was imaginary, what was the corresponding reality?

A mathematician called Carl Gauss helped us understand imaginary numbers. Gauss was born in Brunswick (now Germany) in 1777. Even at the age of 7, his abilities in mathematics were apparent to his teachers. When asked at this young age what the sum of all the integers from 1 to 100 was, he answered immediately. He had intuitively spotted that there are 50 pairs of numbers each adding to 101 (1 + 100, 2 + 99, 3 + 98,... 49 + 52, 50 + 51) so the answer is simply 50 x 101 = 5,050. Pretty good for a 7-year-old.

Above: Portrait of German mathematician Carl Gauss.

Gauss was taught High German, Latin and mathematics. By the time he was 18, he had independently discovered many important mathematical theories and ideas, including the binomial theorem, arithmetic-geometric mean and the prime number theorem. He next studied at Göttingen University where he managed to make some of the most impressive advances in geometry for centuries, including the construction of a new geometric shape, the regular heptadecagon (17-gon), with a ruler and compasses. Just to frighten you, one of the equations he used to construct the shape looked a little like this:

$$\sin(\pi/17)=\sin(180°/17)=\tfrac{1}{8}\sqrt{34-2\sqrt{17}-2\sqrt{2}\sqrt{17-\sqrt{17}}-2\sqrt{68+12\sqrt{17}+2\sqrt{2}(\sqrt{17}-1)\sqrt{17-\sqrt{17}}-16\sqrt{2\sqrt{17}+\sqrt{17}}}}$$

By age 24, Gauss received his doctorate on the fundamental theorem of algebra.

Gauss went on to make many advances in number theory, astronomy, geometry, surveying and physics (especially studies of magnetism).

Gauss' fundamental theorem of algebra

In his doctorate thesis, Gauss formally introduced the notation of a + bi for imaginary numbers. This immediately meant that we could say real numbers are a special type of imaginary numbers, where b equals zero. Gauss also introduced the term complex numbers (which became the most popular name for imaginary numbers) and found the best proof so far for the Fundamental Theorem of Algebra.

Gauss's fundamental theorem of algebra is something of a misnomer since it isn't really about algebra. It's more about being able to say that the field of complex numbers is algebraically closed. This means that given a polynomial equation (something like this: 3 x 2 + 1 = 0), there is a solution in the same field of numbers used for its coefficients. For that equation, the coefficients 3 and 1 are from the field of real numbers, but the solution is √–1/3, which is imaginary, not real. The solution is in a different field of numbers to the coefficients. That's all we need to prove that real numbers are not algebraically closed.

Gauss was one of the first mathematicians to come up with a decent proof that complex numbers are algebraically closed. If you recall, the degree of a polynomial equation means the highest power the x variable is raised to (an equation containing x^2 has degree 2, or containing x^3 has degree 3, or containing x^n has degree n). Gauss's proof showed that every polynomial equation over the field of complex numbers of degree n (where n is higher than 1) has n complex solutions. There will never be a case where an equation of this type does not have a complex solution.

What all this means is that you'll never get stuck trying to find the solutions to an equation when you use complex numbers. So complex numbers are the best numbers to use for tricky equations — a fact used by much of physics today.

With the help of another mathematician named Weber, he even created a telegraph device that could send messages over a distance of 5,000 feet (1,500 m). One of Gauss' last PhD students, named Dedekind, wrote perhaps the best description of what it was like to work with him:

... usually he sat in a comfortable attitude, looking down, slightly stooped, with hands folded above his lap. He spoke quite freely, very clearly, simply and plainly: but when he wanted to emphasize a new viewpoint ... then he lifted his head, turned to one of those sitting next to him, and gazed at him with his beautiful, penetrating blue eyes during the emphatic speech ... If he proceeded from an explanation of principles to the development of mathematical formulas, then he got up, and in a stately very upright posture he wrote on a blackboard beside him in his peculiarly beautiful handwriting: he always succeeded through economy and deliberate arrangement in making do with a rather small space. For numerical examples, on whose careful completion he placed special value, he brought along the requisite data on little slips of paper.

Above: German physicist Wilhelm Eduard Weber, who helped Gauss create his telegraph device.

With the fundamental theorem of algebra proving the usefulness of complex numbers, what began as imaginary concepts rapidly developed into something real. Today they are useful for simplifying calculations in physics and physical systems, but in one area they are more than useful — they are essential. That area is quantum mechanics.

Like Einstein's general theory of relativity, quantum mechanics is another of the great supports underpinning all contemporary physics and technology. While Einstein thought big,

Above: Computer model of a quantum wavefunction trapped in a well. Quantum theory allows a phenomenon known as quantum tunneling, where a particle can "tunnel" and appear in a region impossible in classical physics.

explaining the large physical observable universe around us, quantum mechanics is all about thinking very, very small (see box below).

Like Einstein's general theory of relativity, quantum mechanics is not finished yet. We know this, because its equations don't quite work with the general theory of relativity, which is why Einstein and hundreds of physicists since have attempted to unify the theories to make them all work together. No one has managed it yet, so we need some new geniuses to help us to improve our understanding of how everything works. For now, our understanding of the universe using complex numbers is incomplete, but good enough for our current technology. When a new Einstein comes along, a unified view of physics might lead to some extraordinary new technology, from quantum teleportation to anti-gravity or momentum-nullifying devices.

Quantum mechanics

Quantum mechanics tells us that all energy is quantized — parceled up into packets. Each packet is actually like a particle — so we say that the energy in a beam of light consists of a bunch of photon particles traveling at the speed of light, going through materials such as glass, bouncing off other objects and perhaps hitting your retinas, allowing you to see. There are some great experiments that show that this is exactly what happens. But there are also some great experiments that show that light behaves like a wave as well. That's why there are different colors, light may have different wavelengths. Yet how can light be both particles and waves at the same time?

Quantum mechanics resolves this conundrum by telling us that every particle has a wave function associated with it. That function is a complex function that tells us the probability of a particle being in a particular state. In other words, because of the undecidable nature of such tiny things, a particle might be anywhere, or have any momentum or other measurable property. Its complex wavefunction, combined with some kind of observation, allows us to figure out the real probability of the particle being somewhere specific or having some other specific property. Complex numbers are essential in these calculations. The combination of real part (corresponding to a measurable reality) and complex part (corresponding to an extended reality, not measurable right now) allows a much richer and more complete view of our universe at very small scales. In the words of mathematician, Hadamard:

> The shortest path between two truths in the real domain passes through the complex domain.

Today our ability to calculate and exploit the weird and unpredictable nature of sub-atomic particles is essential in technologies from lasers to microprocessors.

If you're still nervous about complex numbers and think, like Descartes, that they can't ever be real, just remember that a complex number is not very far from a real number. By definition $i^2 = -1$. It's pretty easy to see that $(2i)^2 = -4$ and $i^3 = -i$. And to get perhaps the most intriguing view of i, the value of $i^i = 0.20787957635076190854...$

Complex Visions

It is all very well to write down modified versions of i, or to plot vectors and points corresponding to specific complex numbers, but what of geometry? Using real numbers we are able to describe geometric shapes: solids, surfaces, curves in two, three or even more dimensions. But what is the geometry of complex numbers? Can we use complex numbers to define complex geometric shapes? As Benoit Mandelbrot discovered, the answer is a resounding yes.

Mandelbrot was born in 1924, in Warsaw, Poland. His father sold clothes for a living and his mother was a doctor, but much of the rest of his family were very academic. At the age of 12, Mandelbrot and his family emigrated to France and his uncle, a mathematics professor at the Collège de France, took over his education. His uncle was an enthusiastic follower of Hardy, a pure mathematician and

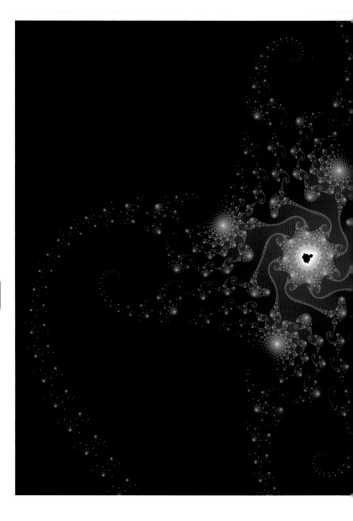

Above: An example of the Mandelbrot set.

extreme pacifist (who believed that applied mathematics could be exploited for weapon development in wartime). The excessive focus on pure math put Mandelbrot off the subject and he took an interest in more applied areas such as geometry. When World War II began, Mandelbrot could no longer attend school regularly and so spent much of his time teaching himself. He later attributed much of his success

The Mandelbrot set

Mandelbrot was interested in the oscillations of complex number series. He thought about a very simple equation:

$$x_{t+1} = x_t^2 + c$$

where x and c were complex numbers and t was time. For any value of c, you could iteratively calculate the value of x_t, increasing t by 1 each time. The equation means, "make the current value of x_t equal to the previous value multiplied by itself, plus the value of c." So (using real numbers to illustrate) if the previous value of x_t is 3, and the value of c is 1, then the current value is 3 x 3 + 1, which is 10. Now if the previous value is 10, then the current value becomes 10 x 10 + 1, which is 21.

Mandelbrot wanted to know which values of c would make the length of the imaginary number stored in x_t stop growing when the equation was applied for an infinite number of times. He discovered that if the length ever went above 2, then it was unbounded — it would grow forever. But for the right imaginary values of c, sometimes the result would simply oscillate between different lengths less than 2. Mandelbrot used his computer to apply the equation many times for different values of c. For each value of c, the computer would stop early if the length of the imaginary number in x_t was 2 or more. If the computer hadn't stopped early for that value of c, a black dot was drawn. The dot was placed at coordinate (m, n) using the numbers from the value of c: $(m + ni)$ where m was varied from —2.4 to 1.34 and n was varied from 1.4 to —1.4, to fill the computer screen.

Mandelbrot expected some kind of geometric shape might appear in the pattern of dots. Perhaps a circle or a square. What he did not expect to see was a "squashed bug" with tendrils and complicated patterns around its edge. To take a closer look, he zoomed in (plotting a smaller range of values of c and magnifying to fill the screen). He discovered more complexity hidden within the patterns, including shapes that looked just like the whole squashed bug. The more he zoomed in to take a closer look, the more complexity was revealed. He quickly realized that the patterns were infinite — no matter how much he zoomed in, there would always be more complexity hiding deeper inside.

to his unconventional and turbulent education, for it enabled him to develop his own insights and intuition in geometry. Mandelbrot studied at École Polytechnique, visited the California Institute of Technology, then took a PhD at the University of Paris. John von Neumann was impressed with the young Mandelbrot and sponsored him to come and work at the Institute for Advanced Study in Princeton. In 1955, two years before von Neumann died, Mandelbrot returned to France. There he met and married Aliette Kagan. But he was unhappy with the style of mathematics at the Centre

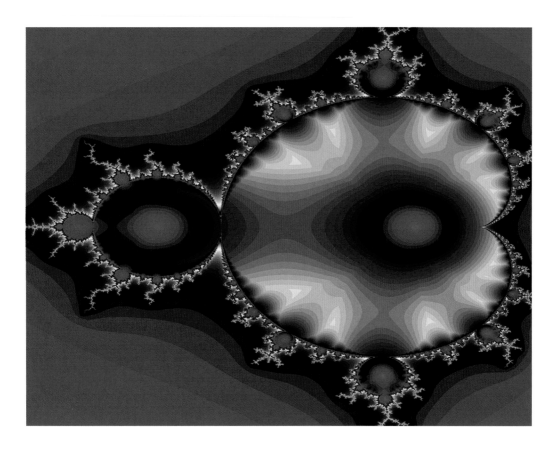

Above: A beautifully elegant example of a Mandelbrot fractal.

National de la Recherche Scientific, and three years later he moved to the United States, joining IBM as a research Fellow at their famous laboratories in Yorktown Heights, New York. He was given access to their high-powered computer facilities and considerable freedom in his research. He used that freedom to perform some of the first ever computer graphics work, plotting the strange shapes that emerged from complex numbers.

Madelbrot named his pattern a fractal, intending to imply the notion of fractional (being able to keep dividing and subdividing the shape forever). Before long he realized that fractal shapes seem to appear everywhere in nature. He compared the infinitely corrugated edge of his fractal to the coastline of an island — the closer you look, the more you see the corrugations of the boundary between land and water. He also compared the self-similarity (the repeated patterns seen at many different scales) with natural shapes such as blood vessels. Today Mandelbrot's fractal is known as the Mandelbrot set (its shape is a visualization of the set of complex numbers that tend to a length less than 2 in the equation). It has become perhaps the most famous and widely seen image ever produced by computers. Yet all you need to do is use a fractal program and zoom in on your own area of the Mandelbrot set and you will be seeing parts of it that no human has ever seen before — there is an infinite amount of it to see.

Since Mandelbrot's discovery, many other varieties of fractals have been discovered. Some lie at the heart of a new area of mathematics known as chaos theory.

The work of Lorenz and Mandelbrot marked the beginning of a new era of mathematics. Instead of relying on analysis and calculation

of the solutions to complicated equations, we could use computers to analyze the equations numerically: put lots of numbers in and see what comes out. The use of computers now enables us to exploit horrifically complicated equations, or vast numbers of simple equations that interact in chaotic or unknown ways. Today we are able to model biological systems and explore how neurons interact with each other in brains, how evolution changes genes, and how our cells interact. This modern form of mathematics is called complexity science, and is leading to new theories on complexity. We now understand that some systems (and particularly biological systems like humans) do not behave in ways that are predictable by conventional mathematics or even chaos theory. When vast numbers of entities interact and dynamically change themselves, new forms of complexity emerge spontaneously, whether during evolution of organisms, flocking of birds, signaling of our immune cells or consciousness

Left: It is not possible to predict the exact movement of a leaky water wheel when water falls from directly above because its motion is chaotic.

Chaos theory

Chaos theory tells us that some systems seem to have random behavior even though they are not random at all. In fact, chaotic systems may have roughly predictable behavior, but they are fundamentally unpredictable in any detail — even if we know the equations that describe those systems.

One example of this is a leaky waterwheel. If the water falls from directly above the water wheel, and each bucket leaks down into buckets below them, then sometimes the wheel will rotate left, and sometimes it will rotate right, depending on which buckets drip into which other buckets below them. Predicting the exact pattern of left and right rotations is not possible — all we can do is predict that there will be a pattern of left and right rotations. Even though there is no randomness in this system (water will always fall down, buckets will always leak at the same rate, the wheel will always rotate when the buckets on one side are heavier than those on the other), its behavior is chaotic.

The pattern of behaviors of chaotic systems may be unpredictable in detail, but the possible behaviors they can have and the transitions from one behavior to another can be calculated, and even drawn. The resulting shapes are known as *strange attractors*, and they are fractal. The pattern may stay the same, but the more you zoom in, the more you see.

One example of the unpredictable nature of chaotic systems is known as the butterfly effect. It means that small perturbations to initial conditions may cause huge and unpredictable effects to a chaotic system. In the waterwheel example, if the wheel was rotated a billionth of a degree more to the left, or if one bucket had a few more molecules of water in it compared to a previous trial run, the resulting pattern of rotations would quickly become totally different. The effect was first noticed by a mathematician called Edward Lorenz who was attempting to use a computer to model the weather in 1961. He decided he would add an option to save the state of the computer model so he could rerun it again at a later date. But his computer saved the results as three-digit numbers, while it worked with six-digit numbers internally, so the saved results were a few decimal places wrong. Conventional mathematics suggested to him that the tiny inaccuracies in the input data should have resulted in tiny and insignificant differences to the predicted results. But when Lorenz reran the model with the very slightly different numbers, the model made completely different weather predictions. Lorenz's model was chaotic — tiny differences to the initial conditions were amplified until they had a huge effect on the predicted weather. The effect became popularized in the media with the idea that the flap of a butterfly's wings on one continent could become amplified and eventually cause completely different weather on another continent, hence the name "butterfly effect." The notion is a bit of a simplification: the behaviors of chaotic systems are attracted to their "strange attractors" and tend to follow unpredictable but similar patterns of behavior. It's sometimes difficult to predict which initial conditions may upset the balance, and which may have no effect at all.

If you were wondering what the fractal strange attractor for the waterwheel and for Lorenz's weather model looked like, they both looked the same:

The shape is called the Lorenz Attractor.

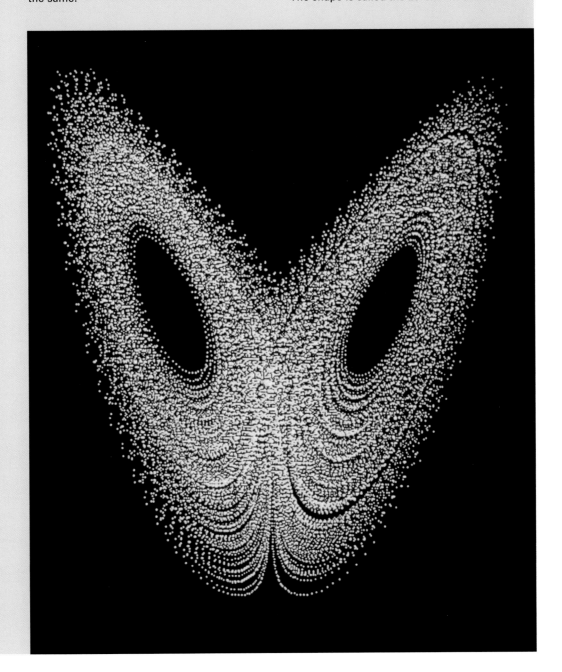

in our brains. By analyzing how and why this complexity arises, we are beginning to understand how other forms of complexity may be controlled. There are a lot of these complex systems to worry about: the spread of disease, the fluctuations of economies, the dynamics of networked computers forming the Internet, predictions of environmental change and even the transmission of knowledge in our cultures. When we understand the numbers that drive complex systems, we will understand the effect our intervention may have in the future (and what we've already done to those systems in the past).

All is Number

The Pythagoreans had a religious fervor about their belief that numbers lay at the heart of the universe. Scientists such as Einstein told us how numbers form the basis of time and space, using equations such as $e=mc^2$. But Euler, the creator of most modern mathematical notation and he who gave up his sight to numbers, provides the final and most elegant combination of numbers in this book. It has been called, "the most profound mathamatical statement ever written," "uncanny and sublime," "filled with cosmic beauty" and "mind-blowing." Physicist Richard Feynman said it was, "the most remarkable formula in mathematics." It looks like this:

$$e^{i\pi}+1 = 0$$

The most remarkable formula in mathematics

Euler discovered the equation during his investigation of imaginary numbers, applying trigonometry to them. In doing so, he was able to show some strange relationships. For example, if i is used to represent an imaginary circular motion, then the following trigonometric relationship holds:

$$e^{i\theta} = \cos \theta + i \sin \theta$$

If we measure angles in radians (angles measured in multiples of π instead of degrees, to make the math easier) then we can describe a semicircular imaginary path with an angle of exactly π (equivalent to 180 degrees).

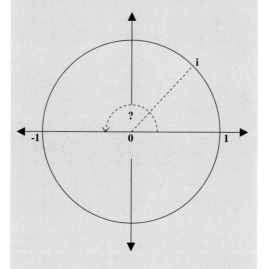

Using the value of π for q, we obtain an extraordinary result:

$$e^{i\pi} = -1$$

Or to write it another way:

$$e^{i\pi} + 1 = 0$$

This simple little mathematical identity is like a light shining in the heart of mathematics. It combines the numbers e, i, π, 1 and 0. It uses arithmetic, calculus, trigonometry, complex analysis and numbers.

Euler's elegant little equation provides us with a subtle clue about the numbers that are woven together in the fabric of reality. They are all aspects of the same thing. Maybe one day we'll discover that all the threads in the tapestry of our universe are linked. Perhaps what we see as different patterns and different numbers are really aspects of a single truth. Is the tapestry of reality made from just one thread? We still have much to learn, but you can be sure of one thing: Our universe is full of patterns that we like to call numbers; while it exists, so will they.

You are made from numbers. So am I. Enjoy the feeling.

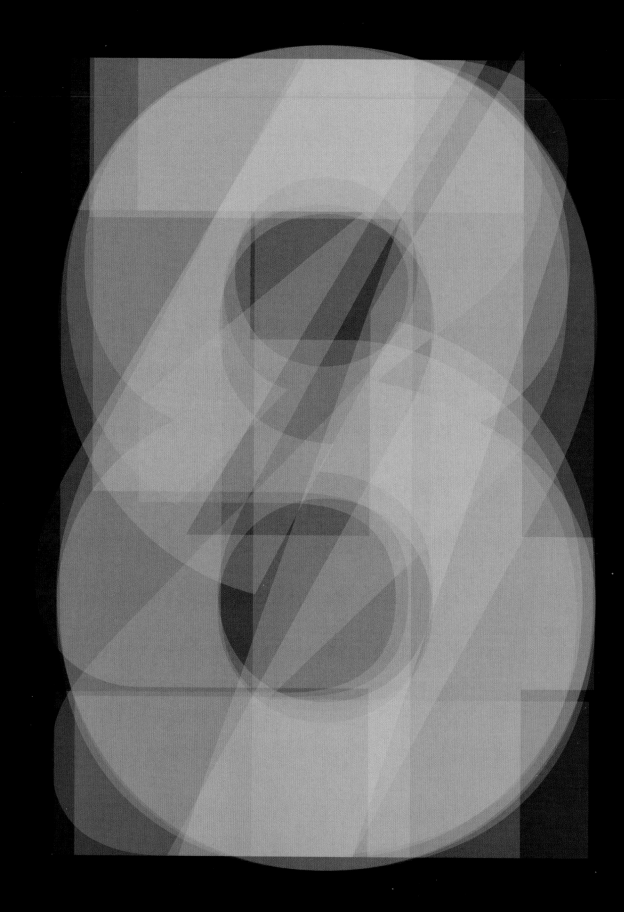

BIBLIOGRAPHY

Web resources
The MacTutor History of Mathematics archive, School of Mathematics and Statistics, University of St. Andrews Scotland: www-history. mcs. st-andrews.ac.uk/history/

Wikipedia: http://en.wikipedia.org/

General
Georges Ifrah. *The Universal History of Numbers vols I, II and III*. London, The Harvill Press, 2000.

Paul J. Nahin. *An Imaginary Tale: The Story of "I" (the Square Root of Minus One)*. Princeton University Press, 1998.

Paul J. Nahin. *Dr. Euler's Fabulous Formula: Cures Many Mathematical Ills*. Princeton University Press, 2006.

Ian Stewart, *Nature's Numbers*. Basic Books, 1995.

Douglas Adams. *The Hitchiker's Guide to the Galaxy*. Pan Macmillan, 1979.

Terry Pratchett, Ian Stewart, Jack Cohen. *Science of Discworld II: The Globe*. Ebury Press, 2002.

Stephen King. *The Dark Tower III: The Waste Lands*. Warner Books, 1992.

Brahmagupta
H. T. Colebrooke. *Algebra, with Arithmetic and Mensuration from the Sanscrit of Brahmagupta and Bhaskara*. 1817.

G. Ifrah. *A Universal History of Numbers: From Prehistory to the Invention of the Computer*. London, 1998.

S. S. Prakash Sarasvati. *A Critical Study of Brahmagupta and His Works: The Most Distinguished Indian Astronomer and Mathematician of the Sixth Century A.D.* Delhi, 1986.

Bhaskara
G. G. Joseph. *The Crest of the Peacock*. London, 1991.

K. S. Patwardhan, S. A. Naimpally and S. L. Singh. *Lilavati of Bhaskaracarya*. Delhi, 2001.

Guillaume de l'Hôpital
J-P Wurtz. *La naissance du calcul différentiel et le problème du statut des infiniment petits: Leibniz et Guillaume de L'Hospital*, in *La mathématique non standard*. Paris, 1989, 13–41.

J. Peiffer. *Le problème de la brachystochrone à travers les relations de Jean I Bernoulli avec L'Hôpital et Varignon*, in *Der Ausbau des Calculus durch Leibniz und die Brüder Bernoulli*. Wiesbaden, 1989, 59–81.

M. C. Solaeche Galera. The L'Hôpital-Bernoulli Controversy (Spanish). *Divulg. Mat.* 1 (1), 1993, 99–104.

Jacob, Johann and Daniel Bernoulli
H. Bernhard, *The Bernoulli Family*, in H. Wussing and W. Arnold. *Biographien bedeutender Mathematiker*. Berlin, 1983.

J. O. Fleckenstein. *Johann und Jacob Bernoulli*. Basel, 1949.

V. A. Nikiforovskii. *The Great Mathematicians Bernoulli (Russian), History of Science and Technology Nauka*. Moscow, 1984.

Aloisius Lilius
The Catholic Encyclopedia, Volume IX. Copyright © 1910 by Robert Appleton Company. www.newadvent.org/cathen/ Online Edition Copyright © 2003 by K. Knight.

Pythagoras
P. Gorman. *Pythagoras, a Life*. 1979.

T. L. Heath. *A History of Greek Mathematics 1*. Oxford, 1931.

Iamblichus. *Life of Pythagoras* (translated into English by T. Taylor). London, 1818.

L. E. Navia. *Pythagoras: An Annotated Bibliography*. New York, 1990.

D. J. O'Meara. *Pythagoras Revived: Mathematics and Philosophy in Late Antiquity*. New York, 1990.

Abu'l Hasan Ahmad ibn Ibrahim Al-Uqlidisi
R. Rashed. *The Development of Arabic Mathematics: Between Arithmetic and Algebra*. London, 1994.

R. Rashed. *Entre arithmétique et algèbre: Recherches sur l'histoire des mathématiques arabes*. Paris, 1984.

A. S. Saidan (trs.). *The Arithmetic of al-Uqlidisi. The Story of Hindu-Arabic Arithmetic as Told in "Kitab al-fusul fial-hisab al-Hindi"* Damascus, A.D. 952/3. Dordrecht-Boston, MA, 1978.

The Arab and his amicable numbers
Martin Gardner. *Mathematical Magic Show*. London, Viking, 1984.

St. Augustine
City of God (great books online). http://books.mirror.org/gb. augustine.html
(accessed July 2006).

Pierre de Fermat
W. W. Rouse Ball. *A Short Account of the History of Mathematics*. 4th edition, 1908.

Leonhard Euler
C. B. Boyer. *The Age of Euler, in A History of Mathematics*. 1968.

R. Thiele. *Leonhard Euler.* Leipzig,1982.

Euclid of Alexandria
C. B. Glavas. *The Place of Euclid in Ancient and Modern Mathematics.* Athens, 1994.

T. L. Heath. *The Thirteen Books of Euclid's Elements* (3 Volumes). New York, 1956.

G. R. Morrow (ed.). *A Commentary on the First Book of Euclid's "Elements."* Princeton, NJ, 1992.

I. Mueller. *Philosophy of Mathematics and Deductive Structure in Euclid's "Elements."* Cambridge, MA-London, 1981.

René Descartes
D. M. Clarke. *Descartes' Philosophy of Science.* 1982.

S. Gaukroger (ed.). *Descartes: Philosophy, Mathematics and Physics.* 1980.

J. F. Scott. *The Scientific Work of René Descartes.* 1987.

W. R. Shea. *The Magic of Numbers and Motion: The Scientific Career of René Descartes.* 1991.

T. Sorell. *Descartes. Past Masters.* New York, 1987.

J. R. Vrooman. *René Descartes: A Biography.* 1970.

Eratosthenes of Cyrene
D. H. Fowler. *The Mathematics of Plato's Academy: A New Reconstruction.* Oxford, 1987.

T. L. Heath. *A History of Greek Mathematics* (2 vols). Oxford, 1921.

Georg Ferdinand Ludwig Philipp Cantor
J. W. Dauben. *Georg Cantor: His Mathematics and Philosophy of the Infinite.* Cambridge, MA, 1979; reprinted 1990.

P. E. Johnson. *A History of Set Theory.* Boston, MA, 1972.

W. Purkert and H. J. Ilgauds. *Georg Cantor 1845–1918.* Basel, 1987.

D. Stander. *Makers of Modern Mathematics: Georg Cantor.* 1989.

Hippocrates of Chios
A. Aaboe. *Episodes from the Early History of Mathematics.* Washington, D.C., 1964.

Iamblichus, *Life of Pythagoras* (translated into English by T. Taylor.) London, 1818.

A. R. Amir-Moéz and J. D. Hamilton. "Hippocrates," J. *Recreational Math.* 7 (2), 1974, 105–107.

Archimedes
A. Aaboe. *Episodes from the Early History of Mathematics.* Washington, D.C., 1964.

R. S. Brumbaugh. *The Philosophers of Greece.* Albany, NY, 1981.

E. J. Dijksterhuis. *Archimedes.* Copenhagen, 1956, and Princeton, NJ, 1987.

T. L. Heath. *A History of Greek Mathematics II.* Oxford, 1931.

Plato
R. S. Brumbaugh. *Plato's Mathematical Imagination: The Mathematical Passages in the Dialogues and Their Interpretation.* Bloomington, IN, 1954.

R. S. Brumbaugh. *The Philosophers of Greece.* Albany, NY, 1981.

G. C. Field. *Plato and His Contemporaries: A Study in Fourth-Century Life and Thought.* 1975.

G. C. Field. *The Philosophy of Plato.* Oxford, 1956.

D. H. Fowler. *The Mathematics of Plato's Academy: A New Reconstruction.* New York, 1990.

F. Lasserre. *The Birth of Mathematics in the Age of Plato.* London, 1964.

J. Moravcsik. *Plato and Platonism: Plato's Conception of Appearance and Reality in Ontology, Epistemology, and Ethics, and its Modern Echoes.* Oxford, 1992.

A. E. Taylor. *Plato, the Man and His Work.* 7th ed. London, 1969.

A. Wedberg. *Plato's Philosophy of Mathematics.* 1977.

Abu Ja'far Muhammad ibn Musa al-Khwarizmi
A. A. al'Daffa. *The Muslim Contribution to Mathematics.* London, 1978.

J. N. Crossley. *The Emergence of Number.* Singapore, 1980.

S. Gandz (ed.). *The Geometry of al-Khwarizmi.* Berlin, 1932.

E. Grant (ed.). *A Source Book in Medieval Science.* Cambridge, 1974.

O. Neugebauer. *The Exact Sciences in Antiquity.* New York, 1969.

R. Rashed. *The Development of Arabic Mathematics: Between Arithmetic and Algebra.* London, 1994.

F. Rosen (trs.). *Muhammad ibn Musa Al-Khwarizmi: Algebra.* London, 1831.

Leonardo da Vinci
M. Clagett. *The Science of Mechanics in the Middle Ages.* Madison, 1959.

K. Clark. *Leonardo da Vinci.* London, 1975.

B. Dibner. *Machines and Weapons, in Leonardo the Inventor.* New York, 1980.

M. Kemp. *Leonardo da Vinci: The Marvelous Works of Nature and Man.* 1981.

R. McLanathan. *Images of the Universe: Leonardo da Vinci: The Artist as Scientist.* 1966.

L. Reti. *The Engineer, in Leonardo the Inventor.* New York, 1980.

V. C. Zubov. *Leonardo da Vinci.* Cambridge, 1968.

Fibonacci
J. Gies and F. Gies. *Leonard of Pisa and the New Mathematics of the Middle Ages.* 1969.

H. Lüneburg. *Leonardi Pisani Liber Abbaci oder Lesevergnügen eines Mathematikers.* Mannheim, 1993

Johannes Kepler
A. Armitage. *John Kepler.* 1966.

C. Baumgardt. *Johannes Kepler: Life and Letters.* New York, NY, 1951.

J. V. Field. *Kepler's Geometrical Cosmology.* Chicago, 1988.

J. Kepler (translated A. M. Duncan, commentary E. J. Aiton). *Mysterium Cosmographicum. The Secret of the Universe.* New York, 1981.

J. Kepler (translated W. Donahue). *Astronomia nova: New Astronomy.* Cambridge, 1992

A Koestler. *The Watershed: A Biography of Johannes Kepler.* 1984.

Simon Stevin
R. Hooykaas and M. G. J. Minnaert (eds.). *Simon Stevin: Science in the Netherlands around 1600.* The Hague, 1970.

D. J. Struik. *The Land of Stevin and Huygens.* Dordrecht-Boston, MA, 1981.

D. J. Struik (ed.). *The Principal Works of Simon Stevin. Vols. IIA, IIB: Mathematics.* Amsterdam, 1958.

K. van Berkel. *The Legacy of Stevin: A Chronological Narrative.* Leiden, 1999.

Gottfried Leibnitz
E. J. Aiton. *Leibniz: A biography.* Bristol- Boston, 1984.

D. Bertoloni Meli. *Equivalence and Priority: Newton versus Leibniz.* New York, 1993.

H. Ishiguro. *Leibniz's Philosophy of Logic and Language.* Cambridge, 1990.

D. Rutherford. *Leibniz and the Rational Order of Nature.* Cambridge, 1995.

R. S. Woolhouse (ed.). *Leibniz: Metaphysics and Philosophy of Science.* London, 1981.

Joseph Jacquard
James Essinger, *Jacquard's Web: How a Hand-Loom Led to the Birth of the Information Age.* Oxford University Press, 2004

Charles Babbage
H. W. Buxton. *Memoir of the Life and Labours of the Late Charles Babbage Esq. F.R.S.* Los Angeles, CA, 1988.

J. M. Dubbey. *The Mathematical Work of Charles Babbage.* Cambridge, 1978.

A. Hyman. *Charles Babbage: Pioneer of the Computer.* Oxford, 1982.

George Boole
D. McHale. *George Boole: His Life and Work.* Dublin, 1985.

G. C. Smith. *The Boole – De Morgan Correspondence, 1842–1864.* New York, 1982.

Bertrand Russell

B. Russell. *The Principles of Mathematics.* Cambridge: At the University Press,1903.

A. J. Ayer. *Bertrand Russell.* 1988.

R. W. Clark. *The Life of Bertrand Russell.* London, J. Cape, 1975.

A. R. Garciadiego (Dantan). *Bertrand Russell and the Origins of the Set-Theoretic "Paradoxes."* Basel, Birkhauser Verlag, 1992.

F. A. Rodriguez-Consuegra. *The Mathematical Philosophy of Bertrand Russell: Origins and Development.* Basel, Birkhauser Verlag, 1991.

R. M. Sainsbury. *Russell.* 1985.

P. A. Schilpp (ed.), *The Philosophy of Bertrand Russell.* 3rd ed. Chicago, Northwestern University, 1944, New York, Harper and Row, 1963.

J. G. Slater. *Bertrand Russell.* Bristol, Thoemmes, 1994.

Kurt Gödel

F. A. Rodriguez-Consuegra (ed.). *Kurt Gödel: Unpublished Philosophical Essays.* Basel, 1995.

H. Wang. *Reflections on Kurt Gödel.* Cambridge, MA, 1987, 2nd ed., 1988.

P. Weingartner and L. Schmetterer (eds.). *Godel Remembered: Salzburg, 10–12 July 1983.* Naples, 1987.

Alan Turing

J. L. Britton, D. C. Ince and P. T. Sanuders (eds.). *Collected Works of A. M. Turing.* 1992.

A. Hodges. *Alan Turing: The Enigma.* 1983.

A. Hodges. *Alan Turing: A Natural Philosopher.* 1997.

S. Turing. *Alan M. Turing.* Cambridge, 1959.

János (John) von Neumann

W. Aspray. *John von Neumann and the Origins of Modern Computing.* Cambridge, 1990.

S. J. Heims. *John von Neumann and Norbert Wiener: From Mathematics to the Technologies of Life and Death.* Cambridge, MA, 1980.

T. Legendi and T. Szentivanyi (eds.). *Leben und Werk von John von Neumann.* Mannheim, 1983.

N. Macrae. *John von Neumann.* New York, 1992.

W. Poundstone. *Prisoner's Dilemma.* Oxford, 1993.

N. A. Vonneuman. *John von Neumann: As Seen by His Brother.* Meadowbrook, PA, 1987.

Claude Shannon

D. Slepian (ed.). *Key Papers in the Development of Information Theory.* New York, Institute of Electrical and Electronics Engineers, Inc, 1974.

N. J. A. Sloane and A. D. Wyner (eds.). *Claude Elwood Shannon: Collected Papers.* New York, 1993.

John Napier

D. J. Bryden. *Napier's Bones: A History and Instruction Manual.* London, 1992.

L. Gladstone-Millar. *John Napier: Logarithm John.* Edinburgh, 2003

C. G. Knott (ed.). *Napier Tercentenary Memorial Volume.* London, 1915.

M. Napier. *Memoirs of John Napier of Merchiston, his Lineage, Life, and Times, with a History of the Invention of Logarithms.* Edinburgh, 1904.

Isaac Newton

Z. Bechler. *Newton's Physics and the Conceptual Structure of the Scientific Revolution.* Dordrecht, 1991.

D. Brewster. *Memoirs of the Life, Writings, and Discoveries of Sir Isaac Newton.* (2 volumes). 1855, reprinted 1965.

S. Chandrasekhar. *Newton's "Principia" for the Common Reader.* New York, 1995.

G. E. Christianson. *In the Presence of the Creator: Isaac Newton and His Times.* 1984.

D. Gjertsen. *The Newton Handbook.* London, 1986.

A. R. Hall. *Isaac Newton, Adventurer in Thought.* Oxford, 1992.

D. B. Meli. *Equivalence and Priority: Newton versus Leibniz. Including Leibniz's Unpublished Manuscripts on the "Principia."* New York, 1993.

R. S. Westfall. *Never at Rest: A Biography of Isaac Newton.* 1990. R. S. Westfall. *The Life of Isaac Newton.* Cambridge, 1993.

August Möbius

J. Fauvel, R. Flood and R. Wilson. *Möbius and his Band.* Oxford, 1993.

Augustus De Morgan

S. E. De Morgan. *Memoir of Augustus De Morgan by his Wife Sophia Elizabeth De Morgan.* London, 1882.

Van Ceulen
D. Huylebrouck. [Ludolph] van Ceulen's [1540–1610] tombstone. *Math. Intelligencer* 17 (4) 1995, 60–61.

Hipparchus
D. R. Dicks. *The Geographical Fragments of Hipparchus.* London, 1960.

O. Neugebauer. *A History of Ancient Mathematical Astronomy.* New York, 1975.

Claudius Ptolemy
A. Aaboe. *On the Tables of Planetary Visibility in the Almagest and the Handy Tables.* 1960.

G. Grasshoff. *The History of Ptolemy's Star Catalogue.* New York, 1990.

R. R. Newton. *The Crime of Claudius Ptolemy.* Baltimore, MD, 1977.

G. J. Toomer (trs.). *Ptolemy's Almagest.* London, 1984.

Galileo Galilei
T. Campanella. *A Defense of Galileo, the Mathematician from Florence.* Notre Dame, IN, 1994.

S. Drake. *Galileo.* Oxford, 1980.

M. A. Finocchiaro. *Galileo and the Art of Reasoning: Rhetorical Foundations of Logic and Scientific Method.* Dordrecht-Boston, MA, 1980.

P. Machamer (ed.). *The Cambridge Companion to Galileo.* Cambridge, 1998.

P. Redondi. *Galileo: Heretic.* Princeton, NJ, 1987.

E. Schmutzer and W. Schütz. *Galileo Galilei* (German). Thun, 1989.

M. Sharratt. *Galileo: Decisive Innovator.* Cambridge, 1994.

Gabriel Mouton
P. Humbert. Les astronomes français de 1610 à 1667, *Bulletin de la Société d'études scientifiques et archéologiques de Draguignan et du Var* 42, 1942, 5–72.

Jérôme Le Français de la Lande (Lalande)
K. Alder. *The Measure of all Things.* London, 2002.

R. Hahn. *The Anatomy of a Scientific Institution: The Paris Academy of Sciences, 1666–1803.* Berkeley, 1971.

Blaise Pascal
D. Adamson. *Blaise Pascal: Mathematician, Physicist and Thinker about God.* Basingstoke, 1995.

F. X. J. Coleman. *Neither Angel nor Beast: The Life and Work of Blaise Pascal.* New York, 1986.

A. W. F. Edwards. *Pascal's Arithmetical Triangle.* New York, 1987.

A. J. Krailsheimer. *Pascal.* 1980.

H. Loeffel. *Blaise Pascal 1623–1662.* Boston- Basel, 1987.

Ole Rømer
Gottfried Kirch. *Astronomie um 1700: Kommentierte Edition des Briefes von Gottfried Kirch an Olaus Römer vom 25. Oktober 1703 (Acta historica stronomiae).* Unknown Binding, 1999.

Centre National de la Recherche Scientifique. *Roemer et la vitesse de la lumière (L'histoire des sciences).* Unknown Binding, 1978.

James Bradley
Rigaud's Memoir prefixed to *Miscellaneous Works and Correspondence of James Bradley, D.D.* Oxford, 1832.

Albert Einstein
M. Beller, J. Renn and R. S. Cohen (eds.). *Einstein in Context.* Cambridge, 1993.

D. Brian. *Einstein – A Life.* New York, 1996.

H. Dukas and B. Hoffmann (eds.). *Albert Einstein: The Human Side. New Glimpses from His Archives.* Princeton, NJ, 1979.

J. Earman, M. Janssen and J. D. Norton (eds.). *The Attraction of Gravitation: New Studies in the History of General Relativity.* Boston, 1993.

D. P. Gribanov. *Albert Einstein's Philosophical Views and the Theory of Relativity "Progress."* Moscow, 1987.

T. Hey and P. Walters. *Einstein's Mirror.* Cambridge, 1997.

G. Holton and Y. Elkana (eds.). *Albert Einstein: Historical and Cultural Perspectives.* Princeton, NJ, 1982.

G. Holton. *Einstein, History, and Other Passions.* Woodbury, NY, 1995.

D. Howard and J. Stachel (eds.). *Einstein and the History of General Relativity.* Boston, MA, 1989.

C. Lánczos. *The Einstein Decade (1905–1915).* New York-London, 1974.

M. White. *Albert Einstein: A Life in Science.* London, 1993.

Zeno of Elea

A. Grunbaum. *Modern Science and Zeno's Paradoxes.* London, 1968.

G. S. Kirk, J. E. Raven and M. Schofield. *The Presocratic Philosophers.* Cambridge, 1983.

W. C. Salmon. *Zeno's Paradoxes.* Indianapolis, IN, 1970.

Aristotle

J. L. Ackrill. *Aristotle the Philosopher.* Oxford, 1981.
D. J. Allan. *The Philosophy of Aristotle.* 1978.

H. G. Apostle. *Aristotle's Philosophy of Mathematics.* Chicago, 1952.

J. Barnes. *Aristotle.* Oxford, 1982.

Z. Bechler. *Aristotle's Theory of Actuality.* Albany, NY, 1995.

W. K. C. Guthrie. *A History of Greek Philosophy, Volume 6, Aristotle: An Encounter.* Cambridge, 1981.

J. P. Lynch. *Aristotle's School: A Study of a Greek Educational Institution.* Berkeley, 1972.

R. Sorabji. *Time, Creation, and the Continuum: Theories in Antiquity and the Early Middle Ages.* 1983.

S. Waterlow. *Nature, Change, and Agency in Aristotle's "Physics."* 1982.

Niccolo Fontana

S. Drake and I. E. Drabkin. *Mechanics in Sixteenth-Century Italy: Selections from Tartaglia, Benedetti, Guido Ubaldo, and Galileo.* 1960.

G. B. Gabrieli. *Nicolo Tartaglia: invenzioni, disfide e sfortune.* Siena, 1986.

Carl Gauss

W. K. Bühler. *Gauss: A Biographical Study.* Berlin, 1981.

T. Hall. *Carl Friedrich Gauss: A Biography.* 1970.

G. M. Rassias (ed.). *The Mathematical Heritage of C. F. Gauss.* Singapore, 1991.

Benoit Mandelbrot

D. J. Albers and G. L. Alexanderson (eds.). *Mathematical People: Profiles and Interviews.* Boston, 1985, 205–226.

P. Clark. Presentation of Professor Benoit Mandelbrot for the Honorary Degree of Doctor of Science. St. Andrews, June 23, 1999.

B. Mandelbrot. Comment j'ai decouvert "les fractales." *La Recherche* 1986, 420–424.

INDEX

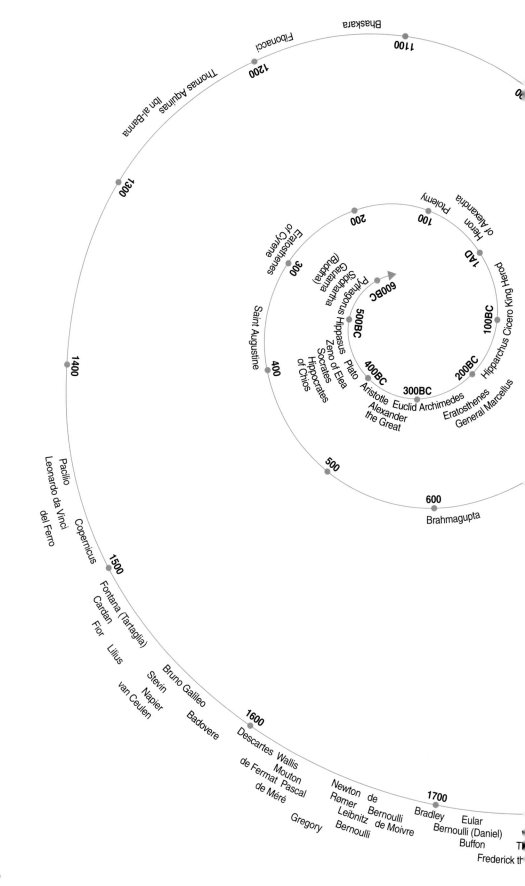

Bhaskara
1100
Fibonacci
1200
Thomas Aquinas Ibn al-Banna
1300
1400
Saint Augustine
Pacilio
Leonardo da Vinci
del Ferro
Copernicus
1500
Fontana (Tartaglia)
Cardan
Fior   Lilius
Bruno Galileo
Stevin
Napier
van Ceulen
Badovere
1600
Descartes Wallis
Mouton
de Fermat Pascal
de Méré
Gregory
Newton de
Rømer Bernoulli
Leibnitz de Moivre
Bernoulli
1700
Bradley Eular
Bernoulli (Daniel)
Buffon
Frederick th

Ptolemy
Heron
of Alexandria
200
100
1AD
Eratosthenes
of Cyrene
300
Herod
Gautama
(Buddha)
Siddhartha
Pythagoras
600BC
Hippasus
Hipparchus Cicero King
500BC
Zeno of Elea
100BC
Socrates
Plato
Hippocrates
400BC
200BC
of Chios
300BC
Eratosthenes
Aristotle Euclid Archimedes
General Marcellus
Alexander
the Great
400
500
600
Brahmagupta

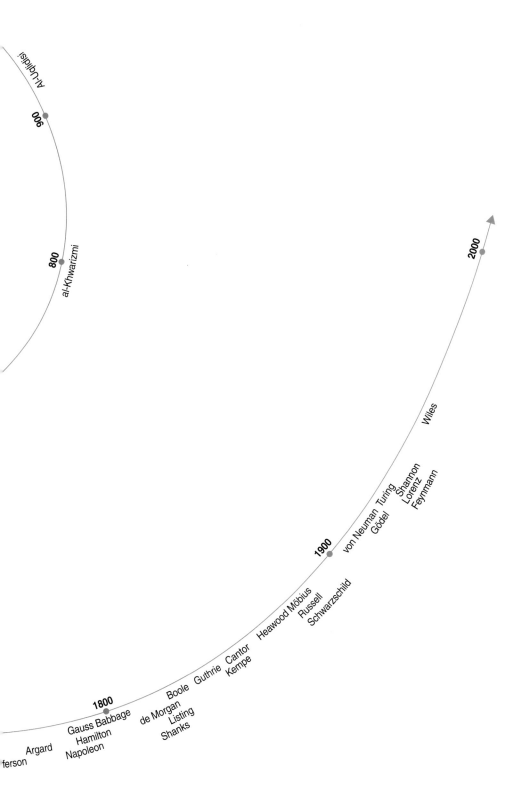

Al-Uqlidisi

900

800 al-Khwarizmi

2000

Wiles

von Neuman Turing
Gödel Shannon
Lorenz
Feynmann

1900

Heawood Möbius
Russell
Schwarzschild

Boole Guthrie Cantor
de Morgan Kempe
Listing
Shanks

1800
Gauss Babbage
Hamilton
Napoleon

Argard

ferson

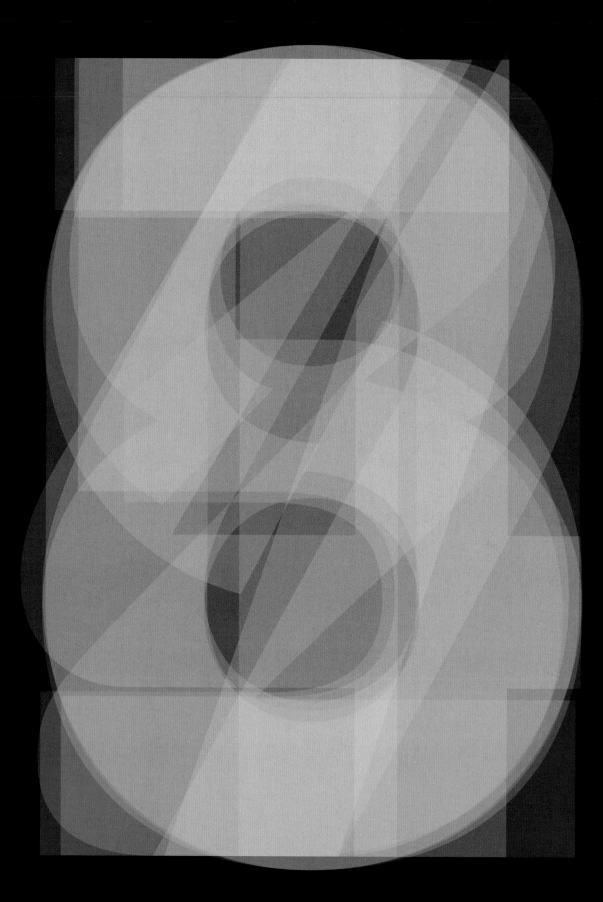

LAST BUT NOT LEAST

Where are the Girls?

A quick note for those perceptive readers who have noticed the lack of females throughout *The Book of Numbers*. This is not for any other reason except for historical accuracy. It is a sad fact that for most of our history, women were not permitted to study at universities. It is also a sad fact, which is true to this day, that female mathematicians and physicists have never been as common as their male counterparts. Perhaps because women prefer more down-to-Earth and practical or cultural pursuits, or perhaps because the education system was geared more towards men, women do not frequently study these topics at university and very rarely go on to devote their lives to research in these areas. This is particularly ironic, for at school, girls are often much better at maths than boys.

There have been exceptions in the past: the Pythagoreans were happy to welcome females as well as males to their sect, Ada Lovelace assisted Babbage in creating the first computer programs,

Florence Nightingale was a statistician as well as a nurse, and often the intelligent wives of many pioneers were helping their husbands with a lot more than dinner. There are also many very successful and eminent female professors today. But the history of numbers has, for much of the last two thousand years, been dominated and coloured by the same male-oriented views as you can see in religious texts of the same eras. (Some intelligent women were even called sorceresses or witches.) If you are female and think the whole thing is unfair, you are right. Rather than complain about inequality, may I encourage you to do something about it, and help change the future history of numbers by becoming involved yourself. The world is a different place today, and education and research positions are positively encouraged for women. With your help, perhaps the next Euler will be Leona, or the next Einstein will be Alberta.

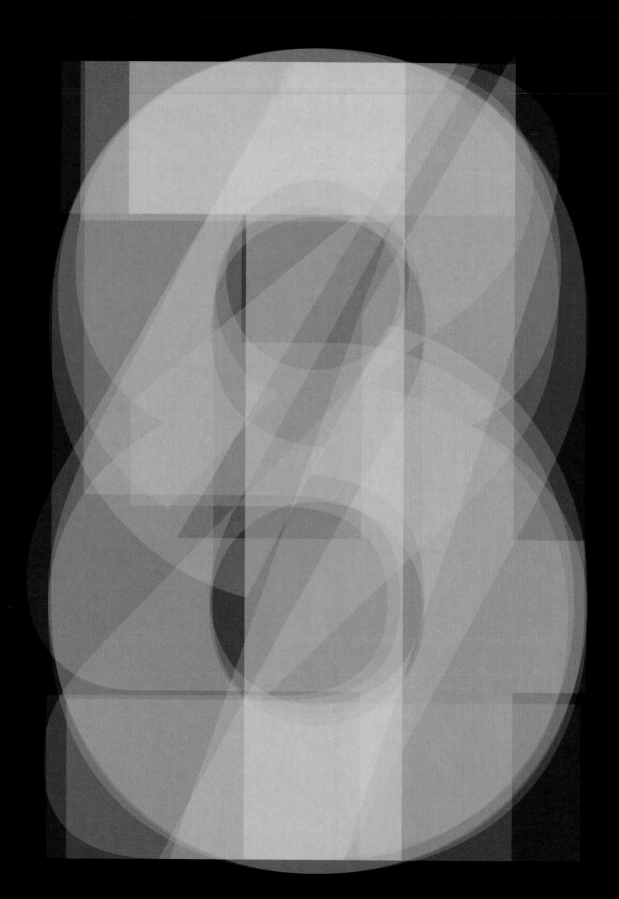

PICTURE ACKNOWLEDGMENTS

AKG images 73, 90, 131, 134,192, 235; Erich Lessing 78; Visioars 31

Alamy Andrew Darrington 13; Classic Image 218; Dale O'Dell 18; Dinodia Images 20; Eddie Gerald; Frappix 248; Israel images 169; Marco Regalia Illustration 245; Mary Evans Picture Library 151, 153; North Wind Picture Archives 80; Peter Arnold Inc. 227; Picturedimensions 119; Israel images 169; Kolvenbach 162; STOCKFOLIO 215; The Print Collector 150; Visual Arts Library (London) 219

Art Archive 186; Bibliothèque des Arts Décoratifs Paris Gianni Dagli Orti 155; Gianni Dagli Orti 14, 15, 16, 25; Galleria degli Uffizi Florence 27

Bridgeman Art Library Bibliotheque Nationale, Paris, France, Lauros Giraudon 45; Academie des Sciences, Paris, France, Giraudon 23; Fitzwilliam Museum, University of Cambridge 38; Galleria dell' Accademia, Venice, Italy, Giraudon 236; Musee de la Ville de Paris, Musee du Petit-Palais, France, Giraudon 171; Louvre, Paris, France, Giraudon 62; Edinburgh University Library 47; Private Collection, Peter Newark American Pictures Private Collection 17; Photo © Christie's Images, 145; Galleria degli Uffizi, Florence, Italy 65; Roy Miles Fine Paintings 177; The Stapleton Collection 42

Corbis 47, 55, 56, 91, 93, 212, 229, 232, 234, 239; Araldo de Luca 19, 41; Archivo Iconografico, S.A. 49, 160,221; ARND Wiegmann/Reuters 210; Bernard Annebicque 129; Bettmann 24, 29,43, 49, 77, 82, 88, 91, 92, 98, 100, 107,114, 115, 118, 122, 123, 125,138, 143, 161, 173, 179, 183, 198, 203, 207, 213, 222, 241 Alan W. Richards 106; Bill Varie 52; Bruno Ehrs 126; Bryan

F. Peterson 202; Digital Art 228; DK Limited, 180; Francis G. Mayer 58, 170; Gianni Dagli Orti 30,168; George B. Diebold 36; Gustavo Tomsich 163; Horace Bristol 32; Hulton-Deutsch Collection 63; Images.com 250; Image Source 190; Joseph Sohm 38; Lester V. Bergman 158; Leonard de Selva 147; Reuters 205; Mark Cooper 184; Matthias Kulka/zefa 159; Michael Nicholson 198; Michael Rosenfeld/dpa 97; Paul Sale Vern Hoffman 11; Paul Souders 194; Sandro Vannini 87; Stapleton Collection 70; Stefano Bianchetti 75, 79; The Art Archive: Alfredo Dagli Orti 74,220; The Gallery Collection 61

Getty altrendo images 249; Image Bank 132; Time and Life pictures 64, 187,102,188; Ian Waldie 105; Marc Romanelli 35; The Italian School 196; Sandra Baker 86

NASA 197,216

Photolibrary 136

Science and Society 26

Science Photo Library 8, 59, 130, 148, 224,226; Astrid & Hanns-Freider Michler 33; CCI Archives 121; Eric Heller 242; Gustoimages 166; George Bernard 112; Jean-Loup Charmet 154, 175; Julien Baum 137; Mark Garlick 134; Prof. E.Lorenz, Peter Arnold Inc. 247; Sandia National Laboratories 34; Science, Industry and Bussiness Library/ New York Public Library 39; Science Source 207; Seymour 159

Superstock Age Fotostock 12

Topfoto 26;Fortean 179; Fortean 179; World History Archive 164,174

Mark Hammonds (illustrations) 83, 93, 175

ACKNOWLEDGMENTS

Thanks to:

Iain MacGregor for the idea.

Gordon Wise for the deals.

Laura Price for being a great editor.

Jenny Doubt for her attention to detail.

Greg Laabs for his statistics from his "pick a random number" website.

Mark Hammonds for his original paintings and designs.

Jools Greensmith for her proofreading and enthusiasm.

The university of St. Andrews for their unbeatable research on the history of mathematics.

Everyone at Cassell Illustrated for helping to produce such a lovely book, and encouraging you, my curious reader, to enjoy it.

And finally (as usual) I would like to thank the cruel and indifferent, yet astonishingly creative process of evolution for providing the inspiration for all of my work. Long may it continue to do so. Where are all the girls?